The Princeton Review®

# Cracking the

# AP®

## BIOLOGY
## EXAM

## 2016 Edition

Kim Magloire and the Staff of The Princeton Review

PrincetonReview.com

Penguin
Random
House

The Princeton Review
24 Prime Parkway, Suite 201
Natick, MA 01760
E-mail: editorialsupport@review.com

Published in the United States by Penguin Random House LLC,
New York, and in Canada by Random House of Canada,
a division of Penguin Random House Ltd., Toronto.

Terms of Service: The Princeton Review Online Companion Tools
("Student Tools") for retail books are available for only the
two most recent editions of that book. Student Tools may be
activated only twice per eligible book purchased for two
consecutive 12-month periods, for a total of 24 months of
access. Activation of Student Tools more than twice per book
is in direct violation of these Terms of Service and may result in
discontinuation of access to Student Tools Services.

ISBN: 978-0-8041-2611-3
eBook ISBN: 978-0-8041-2641-0
ISSN: 1092-0080

AP and Advanced Placement Program are registered trademarks
of the College Board, which does not sponsor or endorse this
product.

The Princeton Review is not affiliated with Princeton University.

Editor: Colleen Day
Production Editor: Liz Rutzel
Production Artist: John E. Stecyk

Printed in the United States of America on partially
recycled paper.

10  9  8  7  6  5  4  3  2

2016 Edition

**Editorial**
Robert Franek, Senior VP, Publisher
Casey Cornelius, VP Content Development
Mary Beth Garrick, Director of Production
Selena Coppock, Managing Editor
Meave Shelton, Senior Editor
Colleen Day, Editor
Sarah Litt, Editor
Aaron Riccio, Editor
Orion McBean, Editorial Assistant

**Random House Publishing Team**
Tom Russell, Publisher
Alison Stoltzfus, Publishing Manager
Melinda Ackell, Associate Managing Editor
Ellen Reed, Production Manager
Kristin Lindner, Production Supervisor
Andrea Lau, Designer

# Acknowledgments

The author would like to give special thanks to her parents, who instilled in her a love of science. She would also like to thank the staff of The Princeton Review and Andrew Taggart for his work on previous editions of this title.

The Princeton Review would like to thank Sarah Woodruff, Deborah Silvestrini, Althea Gladstone, and Bree Porcelli for their hard work on previous revisions to this edition.

The Princeton Review would also like to thank Amanda Benedict for her thorough review of this title and updates to this edition.

# Contents

# Register Your

**1** Go to **PrincetonReview.com/cracking**

**2** You'll see a welcome page where you can register your book using the following ISBN: 9780804126113.

**3** After placing this free order, you'll either be asked to log in or to answer a few simple questions in order to set up a new Princeton Review account.

**4** Finally, click on the "Student Tools" tab located at the top of the screen. It may take an hour or two for your registration to go through, but after that, you're good to go.

If you are experiencing book problems (potential content errors), please contact EditorialSupport@review.com with the full title of the book, its ISBN number (located above), and the page number of the error. Experiencing technical issues? Please e-mail TPRStudentTech@review.com with the following information:

- your full name
- e-mail address used to register the book
- full book title and ISBN
- your computer OS (Mac or PC) and Internet browser (Firefox, Safari, Chrome, etc.)
- description of technical issue

# Book Online!

## Once you've registered, you can...

- Find any late-breaking information released about the AP Biology Exam
- Take a full-length practice PSAT, SAT, and ACT
- Get valuable advice about the college application process, including tips for writing a great essay and where to apply for financial aid
- Sort colleges by whatever you're looking for (such as Best Theater or Dorm), learn more about your top choices, and see how they all rank according to *The Best 380 Colleges*
- Access comprehensive study guides and a variety of printable resources, including bubble sheets and the AP Biology equation and formula tables
- Check to see if there have been any corrections or updates to this edition

### Look For These Icons Throughout The Book

 Online Practice Tests

 Applied Strategies

Proven Techniques

 More Great Books

The **Princeton Review**®
Your Goals. Our Expertise.™

# Part I
# Using This Book to Improve Your AP Score

- Preview: Your Knowledge, Your Expectations
- Your Guide to Using This Book
- How to Begin

## PREVIEW: YOUR KNOWLEDGE, YOUR EXPECTATIONS

Your route to a high score on the AP Biology Exam depends a lot on how you plan to use this book. Respond to the following questions.

1. Rate your level of confidence about your knowledge of the content tested by the AP Biology Exam:

   A. Very confident—I know it all
   B. I'm pretty confident, but there are topics for which I could use help
   C. Not confident—I need quite a bit of support
   D. I'm not sure

2. If you have a goal score in mind, circle your goal score for the AP Biology Exam:

   5      4      3      2      1         I'm not sure yet

3. What do you expect to learn from this book? Circle all that apply to you.

   A. A general overview of the test and what to expect
   B. Strategies for how to approach the test
   C. The content tested by this exam
   D. I'm not sure yet

## YOUR GUIDE TO USING THIS BOOK

This book is organized to provide as much—or as little—support as you need, so you can use this book in whatever way will be most helpful to improving your score on the AP Biology Exam.

- The remainder of **Part I** will provide guidance on how to use this book and help you determine your strengths and weaknesses.

- **Part II** of this book contains Practice Test 1 along with the answers and explanations. (Bubble sheets can be found in the very back of the book for easy tear-out.) We recommend that you take this test before going any further in order to realistically determine:
  - your starting point right now
  - which question types you're ready for and which you might need to practice
  - which content topics you are familiar with and which you will want to carefully review

  Once you have nailed down your strengths and weaknesses with regard to this exam, you can focus your test preparation, build a study plan, and be efficient with your time.

- **Part III** of this book will
  - provide information about the structure, scoring, and content of the AP Biology Exam
  - help you to make a study plan
  - point you towards additional resources

- **Part IV** of this book will explore
  - how to attack multiple-choice questions
  - how to write high-scoring free-response answers
  - how to manage your time to maximize the number of points available to you

- **Part V** of this book covers the content you need for your exam.

- **Part VI** of this book contains Practice Test 2 and its answers and explanations. (Again, bubble sheets can be found in the very back of the book for easy tear-out.) If you skipped Practice Test 1, we recommend that you do both (with at least a day or two between them) so that you can compare your progress between the two. Additionally, this will help to identify any external issues: If you get a certain type of question wrong both times, you probably need to review it. If you only got it wrong once, you may have run out of time or been distracted by something. In either case, this will allow you to focus on the factors that caused the discrepancy in scores and to be as prepared as possible on the day of the test.

You may choose to use some parts of this book over others, or you may work through the entire book. This will depend on your needs and how much time you have. Let's now look how to make this determination.

## HOW TO BEGIN

1. **Take Practice Test 1**

   Before you can decide how to use this book, you need to take a practice test. Doing so will give you insight into your strengths and weaknesses, and the test will also help you make an effective study plan. If you're feeling test-phobic, remind yourself that a practice test is a tool for diagnosing yourself—it's not how well you do that matters but how you use information gleaned from your performance to guide your preparation.

   So, before you read further, take Practice Test 1 starting at page 7 of this book. Be sure to do so in one sitting, following the instructions that appear before the test.

2.  **Check Your Answers**

Using the answer key on page 30, count how many multiple-choice questions you got right and how many you missed. Don't worry about the explanations for now, and don't worry about why you missed questions. We'll get to that soon.

3.  **Reflect on the Test**

After you take your first test, respond to the following questions:

- How much time did you spend on the multiple-choice questions?
- How much time did you spend on each free-response question?
- How many multiple-choice questions did you miss?
- Do you feel you had the knowledge to address the subject matter of the essays?
- Do you feel you wrote well-organized, thoughtful responses to the free-response questions?

4.  **Read Part III of this Book and Complete the Self-Evaluation**

As discussed in the Guide section above, Part III will provide information on how the test is structured and scored. It will also set out areas of content that are tested.

As you read Part III, re-evaluate your answers to the questions above. At the end of Part III, you will revisit and refine the questions you answer above. You will then be able to make a study plan, based on your needs and time available, that will allow you to use this book most effectively.

5.  **Engage with Parts IV and V as Needed**

Notice the word engage. You'll get more out of this book if you use it intentionally than if you read it passively, hoping for an improved score through osmosis.

The strategy chapters in Part IV will help you think about your approach to the question types on this exam. Part IV will open with a reminder to think about how you approach questions now and then close with a reflection section asking you to think about how and/or whether you will change your approach in the future.

The content chapters in Part V are designed to provide a review of the content tested on the AP Biology Exam, including the level of detail you need to know and how the content is tested. You will have the opportunity to assess your mastery of the content of each chapter through test-appropriate questions.

6. **Take Practice Test 2 and Assess Your Performance**

Once you feel you have developed the strategies you need and gained the knowledge you lacked, you should take Practice Test 2, which starts on page 345 of this book. You should do so in one sitting, following the instructions at the beginning of the test.

When you are done, check your answers to the multiple-choice sections. See if a teacher will read your essays and provide feedback.

Once you have taken the test, reflect on what areas you still need to work on, and revisit the chapters in this book that address those deficiencies. Through this type of reflection and engagement, you will continue to improve.

7. **Keep Working**

After you have revisited certain chapters in this book, continue the process of testing, reflection, and engaging with the second practice test in this book. Consider what additional work you need to do and how you will change your strategic approach to different parts of the test.

As we will discuss in Part III, there are other resources available to you, including a wealth of information at AP Central, the official site of the AP Exams. You can continue to explore areas that can stand to improve and engage in those areas right up to the day of the test.

**Want More?**
For extra practice, check out our supplemental title *550 AP Biology Practice Questions*.

# Part II
# Practice Test 1

- Practice Test 1
- Practice Test 1: Answers and Explanations

# Practice Test 1

# AP® Biology Exam

**SECTION I: Multiple-Choice Questions**

## DO NOT OPEN THIS BOOKLET UNTIL YOU ARE TOLD TO DO SO.

### At a Glance

**Total Time**
1 hour and 30 minutes
**Number of Questions**
69
**Percent of Total Score**
50%
**Writing Instrument**
Pencil required

### Instructions

Section I of this examination contains 69 multiple-choice questions. These are broken down into Part A (63 multiple-choice questions) and Part B (6 grid-in questions).

Indicate all of your answers to the multiple-choice questions on the answer sheet. No credit will be given for anything written in this exam booklet, but you may use the booklet for notes or scratch work. After you have decided which of the suggested answers is best, completely fill in the corresponding oval on the answer sheet. Give only one answer to each question. If you change an answer, be sure that the previous mark is erased completely. Here is a sample question and answer.

Sample Question                    Sample Answer

Chicago is a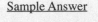

(A)     state
(B)     city
(C)     country
(D)     continent

Use your time effectively, working as quickly as you can without losing accuracy. Do not spend too much time on any one question. Go on to other questions and come back to the ones you have not answered if you have time. It is not expected that everyone will know the answers to all the multiple-choice questions.

**About Guessing**

Many candidates wonder whether or not to guess the answers to questions about which they are not certain. Multiple choice scores are based on the number of questions answered correctly. Points are not deducted for incorrect answers, and no points are awarded for unanswered questions. Because points are not deducted for incorrect answers, you are encouraged to answer all multiple-choice questions. On any questions you do not know the answer to, you should eliminate as many choices as you can, and then select the best answer among the remaining choices.

**This page intentionally left blank.**

**BIOLOGY**

**SECTION I**

**69 Questions**

**Time—90 minutes**

<u>Directions:</u> Each of the questions or incomplete statements below is followed by four suggested answers or completions. Select the one that is best in each case and then fill in the corresponding oval on the answer sheet.

1. The resting membrane potential depends on which of the following?

   I. Active transport

   II. Selective permeability

   III. Differential distribution of ions across the axonal membrane

   (A) III only
   (B) I and II only
   (C) II and III only
   (D) I, II, and III

2. The Krebs cycle in humans occurs in the

   (A) mitochondrial matrix
   (B) inner mitochondrial membrane
   (C) outer mitochondrial membrane
   (D) intermembrane space

3. A heterotroph

   (A) obtains its energy from sunlight, harnessed by pigments
   (B) obtains its energy by oxidizing organic molecules
   (C) makes organic molecules from $CO_2$
   (D) obtains its energy by consuming exclusively autotrophs

4. Regarding meiosis and mitosis, one difference between the two forms of cellular reproduction is that in meiosis

   (A) there is one round of cell division, whereas in mitosis there are two rounds of cell division
   (B) separation of sister chromatids occurs during the second division, whereas in mitosis separation of sister chromatids occurs during the first division
   (C) chromosomes are replicated during interphase, whereas in mitosis chromosomes are replicated during prophase
   (D) spindle fibers form during prophase, whereas in mitosis the spindle fibers form during metaphase

5. A feature of amino acids NOT found in carbohydrates is the presence of

   (A) carbon atoms
   (B) oxygen atoms
   (C) nitrogen atoms
   (D) hydrogen atoms

6. Which of the following is NOT a characteristic of bacteria?

   (A) Circular double-stranded DNA
   (B) Membrane-bound cellular organelles
   (C) Plasma membrane consisting of lipids and proteins
   (D) Ribosomes that synthesize polypeptides

**GO ON TO THE NEXT PAGE.**

7. Which of the following best explains why a population is described as the evolutionary unit?

   (A) Genetic changes can only occur at the population level.
   (B) The gene pool in a population remains fixed over time.
   (C) Natural selection affects individuals, not populations.
   (D) Individuals cannot evolve, but populations can.

8. The endocrine system maintains homeostasis using many feedback mechanisms. Which of the following is an example of positive feedback?

   (A) Infant suckling causes a mother's brain to release oxytocin, which in turn stimulates milk production.
   (B) An enzyme is allosterically inhibited by the product of the reaction it catalyzes.
   (C) When ATP is abundant the rate of glycolysis decreases.
   (D) When blood sugar levels decrease to normal after a meal, insulin is no longer secreted.

9. A scientist carries out a cross between two guinea pigs, both of which have black coats. Black hair coat is dominant over white hair coat. Three quarters of the offspring have black coats, and one quarter have white coats. The genotypes of the parents were most likely

   (A) bb × bb
   (B) Bb × Bb
   (C) Bb × bb
   (D) BB × Bb

10. A large island is devastated by a volcanic eruption. Most of the horses die except for the heaviest males and heaviest females of the group. They survive, reproduce, and perpetuate the population. Since weight is highly heritable and the distribution of weights approximates a binomial distribution, the offspring of the next generation would be expected to have

    (A) a higher mean weight compared with their parents
    (B) a lower mean weight compared with their parents
    (C) the same mean weight as members of the original population
    (D) a higher mean weight compared with members of the original population

11. All of the following play a role in morphogenesis EXCEPT

    (A) apoptosis
    (B) homeotic genes
    (C) operons
    (D) inductive effects

12. During the period when life is believed to have begun, the atmosphere on primitive Earth contained abundant amounts of all the following gases EXCEPT

    (A) oxygen
    (B) hydrogen
    (C) ammonia
    (D) methane

13. Villi and microvilli are present in the small intestine and aid in reabsorption by

    (A) increasing the surface area of the small intestine
    (B) decreasing the surface area of the small intestine
    (C) making the small intestine more hydrophilic
    (D) making the small intestine more hydrophobic

14. Which of the following does NOT take place in the small intestine?

    (A) Pancreatic lipase breaks down fats to fatty acids and glycerol.
    (B) Pepsin breaks down proteins to amino acids.
    (C) Pancreatic amylase breaks down carbohydrates into simple sugars.
    (D) Bile emulsifies fats into smaller fat particles.

15. In animal cells, which of the following represents the most likely pathway that a secretory protein takes as it is synthesized in a cell?

    (A) Plasma membrane–Golgi apparatus–ribosome–secretory vesicle–rough ER
    (B) Ribosome–Golgi apparatus–rough ER–secretory vesicle–plasma membrane
    (C) Plasma membrane–Golgi apparatus–ribosome–secretory vesicle–rough ER
    (D) Ribosome–rough ER–Golgi apparatus–secretory vesicle–plasma membrane

**GO ON TO THE NEXT PAGE.**

16. All of the following statements are correct regarding alleles EXCEPT

    (A) alleles are alternative forms of the same gene
    (B) alleles are found on corresponding loci of homologous chromosomes
    (C) a gene can have more than two alleles
    (D) an individual with two identical alleles is said to be heterozygous with respect to that gene

17. Once a plasmid has incorporated specific genes, such as the gene coding for the antibiotic ampicillin, into its genome, the plasmid may be cloned by

    (A) inserting it into a virus to generate multiple copies
    (B) treating it with a restriction enzyme in order to cut the molecule into small pieces
    (C) inserting it into a suitable bacterium in order to produce multiple copies
    (D) running it on a gel electrophoresis in order to determine the size of the gene of interest

18. Although mutations occur at a regular and predictable rate, which of the following statements is the LEAST likely reason the frequency of mutation appears to be low?

    (A) Some mutations produce alleles that are recessive and may not be expressed.
    (B) Some undesirable phenotypic traits may be prevented from reproducing.
    (C) Some mutations cause such drastic phenotypic changes that they are removed from the gene pool.
    (D) The predictable rate of mutation results in ongoing variability in a gene pool.

19. Which of the following adaptive features would most likely be found in an animal living in a hot arid environment?

    (A) Long loops of Henle to maximize water reabsorption
    (B) Storage of water in fatty tissues
    (C) Large ears to aid in heat dispersion
    (D) Short loops of Henle to maximize water secretion

20. Which of the following best accounts for the ability of legumes to grow well in nitrogen-poor soils?

    (A) These plants make their own proteins.
    (B) These plants have a mutualistic relationship with nitrogen-fixing bacteria.
    (C) These plants are capable of directly converting nitrogen gas into nitrates.
    (D) These plants do not require nitrogen to make plant proteins.

21. Which of the following is most correct concerning cell differentiation in vertebrates?

    (A) Cells in different tissues contain different sets of genes, leading to structural and functional differences.
    (B) Differences in the timing and expression levels of different genes leads to structural and functional differences.
    (C) Differences in the reading frame of mRNA leads to structural and functional differences.
    (D) Differences between tissues result from spontaneous morphogenesis.

Question 22 refers to the diagram below.

22. Which of the following chambers or vessels carry deoxygenated blood in the human heart?

    (A) 4 only
    (B) 1 and 2 only
    (C) 5 only
    (D) 1, 2, and 4

**GO ON TO THE NEXT PAGE.**

23. In chick embryos, the extraembryonic membrane that provides nourishment to the fetus is the

    (A) amnion
    (B) chorion
    (C) placenta
    (D) egg yolk

24. Some strains of viruses can change normal mammalian cells into cancer cells in vitro. This transformation of the mammalian cell is usually associated with the

    (A) formation of a pilus between the mammalian cell and the virus
    (B) incorporation of the viral genome into the mammalian cell's nuclear DNA
    (C) conversion of the host's genome into the viral DNA
    (D) release of spores into the mammalian cell

25. The major difference between cartilage and bone is that cartilage

    (A) is a part of the skeletal system
    (B) is composed of collagen and salts
    (C) lacks blood vessels and nerves
    (D) secretes a matrix

26. All of the following are examples of events that can prevent interspecific breeding EXCEPT

    (A) the potential mates experience geographic isolation
    (B) the potential mates experience behavioral isolation
    (C) the potential mates have different courtship rituals
    (D) the potential mates have similar breeding seasons

27. Which of the following is NOT a characteristic of asexual reproduction in animals?

    (A) Daughter cells have the same number of chromosomes as the parent cell.
    (B) Daughter cells are identical to the parent cell.
    (C) The parent cell produces diploid cells.
    (D) The daughter cells fuse to form a zygote.

28. Which of the following is the correct characteristic of arteries?

    (A) They are thin-walled blood vessels.
    (B) They contain valves that prevent backflow.
    (C) They always carry oxygenated blood.
    (D) They carry blood away from the heart.

29. Transpiration is a result of special properties of water. The special properties of water include all of the following EXCEPT

    (A) cohesion
    (B) adhesion
    (C) capillary action
    (D) hydrophobicity

30. Crossing-over occurs during which of the following phases in meiosis?

    (A) Prophase I
    (B) Metaphase I
    (C) Anaphase I
    (D) Prophase II

31. Which of the following about meiosis is NOT true?

    (A) Meiosis produces two haploid gametes.
    (B) Homologous chromosomes join during synapsis.
    (C) Sister chromatids separate during meiosis I.
    (D) Crossing-over increases genetic variation in gametes.

32. A plant grows in the opposite direction of the gravitational force. This is an example of

    (A) positive thignotropism
    (B) negative phototropism
    (C) positive phototropism
    (D) negative gravitropism

**GO ON TO THE NEXT PAGE.**

33. In most ecosystems, net primary productivity is important because it represents the

    (A) energy available to producers
    (B) total solar energy converted to chemical energy by producers
    (C) biomass of all producers
    (D) energy available to heterotrophs

34. Hawkmoths are insects that are similar in appearance and behavior to hummingbirds. Which of the following is LEAST valid?

    (A) These organisms are examples of convergent evolution.
    (B) These organisms were subjected to similar environmental conditions.
    (C) These organisms are genetically related to each other.
    (D) These organisms have analogous structures.

35. Which of the following describes a symbiotic relationship?

    (A) A tapeworm feeds off of its host's nutrients causing the host to lose large amounts of weight.
    (B) Certain plants grow on trees in order to gain access to sunlight, not affecting the tree.
    (C) Remora fish eat parasites off of sharks; the sharks stay free of parasites, and the remora fish are protected from predators.
    (D) Meerkats sound alarm calls to warn other meerkats of predators.

36. Destruction of all beta cells in the pancreas will cause which of the following to occur?

    (A) Glucagon secretion will stop and blood glucose levels will increase.
    (B) Glucagon secretion will stop and blood glucose levels will decrease.
    (C) Glucagon secretion will stop and digestive enzymes will be secreted.
    (D) Insulin secretion will stop and blood glucose levels will increase.

37. All of the following are stimulated by the sympathetic nervous system EXCEPT

    (A) dilation of the pupil of the eye
    (B) constriction of blood vessels
    (C) increased secretion of the sweat glands
    (D) increased peristalsis in the gastrointestinal tract

38. The calypso orchid, *Calypso bulbosa*, grows in close association with mycorrhizae fungi. The fungi penetrate the roots of the flower and take advantage of the plant's food resources. The fungi concentrate rare minerals, such as phosphates, in the roots and make them readily accessible to the orchid. This situation is an example of

    (A) parasitism
    (B) commensalism
    (C) mutualism
    (D) endosymbiosis

39. Which of the following are characteristics of both bacteria and fungi?

    (A) Cell wall, DNA, and plasma membrane
    (B) Nucleus, organelles, and unicellularity
    (C) Plasma membrane, multicellularity, and Golgi apparatus
    (D) Cell wall, unicellularity, and mitochondria

40. A sustained decrease in circulating $Ca^{2+}$ levels might be caused by decreased levels of which of the following substances?

    (A) Growth hormone
    (B) Parathyroid hormone
    (C) Thyroid hormone
    (D) Calcitonin

41. The synthesis of new proteins necessary for lactose utilization by the bacterium *E. coli* using the *lac* operon is regulated

    (A) by the synthesis of additional ribosomes
    (B) at the transcription stage
    (C) at the translation stage
    (D) by differential replication of the DNA that codes for lactose-utilizing mechanisms

42. Which of the following statements about trypsin is NOT true?

    (A) It is an organic compound made of proteins.
    (B) It is a catalyst that alters the rate of a reaction.
    (C) It is operative over a wide pH range.
    (D) The rate of catalysis is affected by the concentration of substrate.

**GO ON TO THE NEXT PAGE.**

43. In DNA replication, which of the following does NOT occur?

    (A) Helicase unwinds the double helix.
    (B) DNA ligase links the Okazaki fragments.
    (C) RNA polymerase is used to elongate both chains of the helix.
    (D) DNA strands grow in the 5' to 3' direction.

44. A change in a neuron membrane potential from +50 millivolts to –70 millivolts is considered

    (A) depolarization
    (B) repolarization
    (C) hyperpolarization
    (D) an action potential

45. The energy given up by electrons as they move through the electron transport chain is used to

    (A) break down glucose
    (B) make glucose
    (C) produce ATP
    (D) make NADH

46. If a photosynthesizing plant began to release $^{18}O_2$ instead of normal oxygen, one could most reasonably conclude that the plant had been supplied with

    (A) $H_2O$ containing radioactive oxygen
    (B) $CO_2$ containing radioactive oxygen
    (C) $C_6H_{12}O_6$ containing radioactive oxygen
    (D) $NO_2$ containing radioactive oxygen

47. All of the following statements describe the unique characteristics of water EXCEPT

    (A) it is a polar solvent
    (B) it forms hydrogen bonds with disaccharides
    (C) it can dissociate into hydrogen ions and hydroxide ions
    (D) it is a hydrophobic solvent

48. Chemical substances released by organisms that elicit a physiological or behavioral response in other members of the same species are known as

    (A) auxins
    (B) hormones
    (C) pheromones
    (D) enzymes

49. Homologous structures are often cited as evidence for the process of natural selection. All of the following are examples of homologous structures EXCEPT

    (A) the wings of a bird and the wings of a bat
    (B) the flippers of a whale and the arms of a man
    (C) the pectoral fins of a porpoise and the flippers of a seal
    (D) the forelegs of an insect and the forelimbs of a dog

50. The sliding action in the myofibril of skeletal muscle contraction requires which of the following?

    I. $Ca^{2+}$

    II. ATP

    III. Actin

    (A) I only
    (B) II only
    (C) II and III
    (D) I, II, and III

51. Certain populations of finches have long been isolated on the Galapagos Islands off the western coast of South America. Compared with the larger stock population of mainland finches, these separate populations exhibit far greater variation over a wider range of species. The variation among these numerous finch species is the result of

    (A) convergent evolution
    (B) divergent evolution
    (C) disruptive selection
    (D) stabilizing selection

52. Which of the following contributes the MOST to genetic variability in a population?

    (A) Sporulation
    (B) Binary fission
    (C) Vegetative propagation
    (D) Mutation

**GO ON TO THE NEXT PAGE.**

Directions: Each group of questions below concerns an experimental or laboratory situation or data. In each case, first study the description of the situation or data. Then choose the one best answer to each question following it and fill in the corresponding oval on the answer sheet.

Questions 53–55 refer to the following information and table.

A marine ecosystem was sampled in order to determine its food chain. The results of the study are shown below.

| Type of Organism | Number of Organisms |
|---|---|
| Shark | 2 |
| Small crustaceans | 400 |
| Mackerel | 20 |
| Phytoplankton | 1,000 |
| Herring | 100 |

53. Which of the following organisms in this population are secondary consumers?

(A) Sharks
(B) Mackerels
(C) Herrings
(D) Small crustaceans

54. Which of the following organisms has the largest biomass in this food chain?

(A) Phytoplanktons
(B) Mackerels
(C) Herrings
(D) Sharks

55. If the herring population is reduced by predation, which of the following is most likely to occur in this aquatic ecosystem?

(A) The mackerels will be the largest predator in the ecosystem.
(B) The small crustacean population will be greatly reduced.
(C) The phytoplankton population will be reduced over the next year.
(D) The small crustaceans will become extinct.

**GO ON TO THE NEXT PAGE.**

Questions 56–58 refer to the following information and diagram.

To understand the workings of neurons, an experiment was conducted to study the neural pathway of a reflex arc in frogs. A diagram of a reflex arc is given below.

Neural Action

56. Which of the following represents the correct pathway taken by a nerve impulse as it travels from the spinal cord to effector cells?

(A) 1-2-3-4
(B) 6-5-4-3
(C) 2-3-4-5
(D) 4-5-6-7

57. The brain of the frog is destroyed. A piece of acid-soaked paper is applied to the frog's skin. Every time the piece of paper is placed on its skin, one leg moves upward. Which of the following conclusions is best supported by the experiment?

(A) Reflex actions are not automatic.
(B) Some reflex actions can be inhibited or facilitated.
(C) All behaviors in frogs are primarily reflex responses.
(D) This reflex action bypasses the brain.

58. A nerve impulse requires the release of neurotransmitters at the axonal bulb of a presynaptic neuron. Which of the following best explains the purpose of neurotransmitters, such as acetylcholine?

(A) They speed up the nerve conduction in a neuron.
(B) They open the sodium channels in the axonal membrane.
(C) They excite or inhibit the postsynaptic neuron.
(D) They open the potassium channels in the axonal membrane.

**GO ON TO THE NEXT PAGE.**

Questions 59–61 refer to the figure and chart below.

AUGCCACUAGCACGU    mRNA

5 NACGG    3

| Met | Pro | Leu | Ala | Arg | Protein |

Methionine  Proline  Leucine  Alanine  Arginine

### Formation of a Protein

| | The Genetic Code: Codons of mRNA that Specify a Given Amino Acid | | | |
|---|---|---|---|---|
| First Position (5' end) | Second Position | Third Position (3' end) | | |
| | | U | C | A | G |
| U | U | UUU UUC **Phenylalanine** | UUA UUG **Lucine** | | |
| | C | UCU UCC UCA UCG **Serine** | | | |
| | A | UAU UAC **Tytrosine** | UAA UAG | | |
| | G | UGU UGC **Cysteine** | UGA | UGG **Tryptophan** | |
| C | U | CUU CUC CUA CUG **Leucine** | | | |
| | C | CCU CCC CCA CCG **Proline** | | | |
| | A | CAU CAC **Histidine** | CAA CAG **Glutamine** | | |
| | G | CGU CGC CGA CGG **Arginine** | | | |
| A | U | AUU AUC AUA **Isoleucine** | AUG | | |
| | C | ACU ACC ACA ACG **Threonine** | | | |
| | A | AAU AAC **Asparagine** | AAA AAG **Lysine** | | |
| | G | AGU AGC **Serine** | AGA AGG **Arginine** | | |
| G | U | GUU GUC GUA GUG **Valine** | | | |
| | C | GCU GCC GCA GCG | | | |
| | A | GAU GAC **Aspartic Acid** | GAA GAG **Glutamic acid** | | |
| | G | GGU GGC GGA GGG **Glycine** | | | |

59. Which of the following DNA strands would serve as a template for the amino acid sequence shown above?

(A) 3'-ATGCGACCAGCACGT-5'
(B) 3'-AUGCCACUAGCACGU-5'
(C) 3'-TACGGTGATCGTGCA-5'
(D) 3'-UACGGUGAUCGUGCA-5'

60. If a mutation occurs in which uracil is deleted from the messenger RNA after methionine is translated, which of the following represents the resulting amino acid sequence?

(A) Serine–histidine–serine–threonine
(B) Methionine–proline–glutamine–histidine
(C) Methionine–proline–leucine–alanine–arginine
(D) Methionine–proline–alanine–arginine–arginine

61. The mRNA above was found to be much smaller than the original mRNA synthesized in the nucleus. This is due to the

(A) addition of a poly(A) tail to the mRNA molecule
(B) addition of a cap to the mRNA molecule
(C) excision of exons from the mRNA molecule
(D) excision of introns from the mRNA molecule

**GO ON TO THE NEXT PAGE.**

Questions 62–63 refer to the following information.

A scientist studies the storage and distribution of oxygen in humans and Weddell seals to examine the physiological adaptations that permit seals to descend to great depths and stay submerged for extended periods. The figure below depicts the oxygen storage in both organisms.

Human (70 kilograms)

Muscle 13%
Blood 51%
Lungs 36%

Total oxygen store: 1.95 liters

Weddell seal (450 kilograms)

Muscle 25%
Blood 70%
Lungs 5%

Total oxygen store: 25.9 liters

62. Compared with humans, approximately how many liters of oxygen does the Weddell seal store per kilogram of body weight?

(A) The same amount of oxygen
(B) Twice the amount of oxygen
(C) Three times the amount of oxygen
(D) Five times the amount of oxygen

63. During a dive, a Weddell seal's blood flow to the abdominal organs is shut off and oxygen-rich blood is diverted to the eyes, brain, and spinal cord. Which of the following is the most likely reason for this adaptation?

(A) To increase the number of red blood cells in the nervous system
(B) To increase the amount of oxygen reaching the skeletomuscular system
(C) To increase the amount of oxygen reaching the central nervous system
(D) To increase the oxygen concentration in the lungs

**GO ON TO THE NEXT PAGE.**

Directions: Part B consists of questions requiring numeric answers. Calculate the correct answer for each question.

64. An experiment was conducted to observe the light-absorbing properties of chlorophylls and carotenoids using a spectrophotometer. The pigments were first extracted and dissolved in a solution. They were then illuminated with pure light of different wavelengths to detect which wavelengths were absorbed by the solution. The results are presented in the absorption spectrum below.

Absorption Spectrum for Green Plants

.......... Chlorophyll *a*
———— Chlorophyll *b*
‑ ‑ ‑ ‑ Carotenoids

At approximately what wavelength does chlorophyll *a* maximally absorb light?

65. A woman with blood genotype $I^A i$ and a man with blood genotype $I^B i$ have two children, both type AB. What is the probability that a third child will be blood type AB?

66. The trophic level efficiency of large herbivores such as elks is frequently only about 5 percent. In tons, what volume of plants would be required to maintain 24,000 lbs of elk?

67. If the genotype frequencies of an insect population are AA = 0.49, Aa = 0.42, and aa = 0.09, what is the gene frequency of the recessive allele?

$$\begin{array}{r} 24 \\ \times 20 \\ \hline 480 \end{array}$$

**GO ON TO THE NEXT PAGE.**

Question 68 refers to the diagram below.

Family Tree for Color Blindness

69. The loss of water by evaporation from the leaf openings is known as transpiration. The transpiration rates of various plants are shown below.

Transpiration Rates for Plants

68. Based on the pedigree above, what is the probability that a male child born to individuals 6 and 7 will be color-blind?

How many liters of water per week are lost by a coconut palm?

**END OF SECTION I**

This page intentionally left blank.

# BIOLOGY
## SECTION II
**8 Questions**
**Planning Time—10 minutes**
**Writing Time—80 minutes**

Directions: Questions 1 and 2 are long free-response questions that should require about 22 minutes each to answer and are worth 10 points each. Questions 3 through 8 are short free-response questions that should require about 6 minutes each to answer. Questions 3 through 5 are worth 4 points each, and questions 6 through 8 are worth 3 points each.

Read each question carefully and completely. Write your response in the space provided following each question. Only material written in the space provided will be scored. Answers must be written out in paragraph form. Outlines, bulleted lists, or diagrams alone are not acceptable.

1. Cell size is limited by the surface area to volume ratio of the cell membrane.

    (a) **Discuss** why cell size is limited by this ratio.

    (b) **Describe** two adaptations that increase surface area in organisms.

    (c) **Describe** the processes by which small polar and small nonpolar molecules can cross cell membranes according to their concentration gradients, and give an example of each type of molecule.

2. Sickle-cell anemia is a genetic disorder caused by the abnormal gene for hemoglobin S. A single substitution occurs in which glutamic acid is substituted for valine in the sixth position of the hemoglobin molecule. This change reduces hemoglobin's ability to carry oxygen.

    (a) **Discuss** the process by which mutation occurs in base substitution.

    (b) Biologists used gel electrophoresis to initially identify the mutant gene. **Explain** how gel electrophoresis could be applied to the identification of the gene mutation. **Discuss** the use of restriction enzymes.

    (c) Hemoglobin S is transmitted as a simple Mendelian allele. **Describe** the outcome if a female who does not carry the abnormal allele mates with a male homozygous for the disease. **Include** a Punnett square and phenotypic and genotypic ratios.

3. The cell membrane is an important structural feature of a nerve cell.

    (a) **Discuss** what ions and concentration are associated with the resting state of a neuron.

    (b) **Describe** the role of membranes in the conduction of a nerve impulse.

4. **Discuss** the Krebs cycle, the electron transport chain, and oxidative phosphorylation.

    (a) **Explain** why these steps are considered aerobic processes.

    (b) **Discuss** the location at which **each** stage occurs.

**GO ON TO THE NEXT PAGE.**

5. **Describe** three main differences between meiosis and mitosis.

6. **Define** homologous structures and give an example.

7. **Describe** the three ways that genetic information is transmitted laterally between bacteria.

8. **Describe** why viruses are typically not considered to be alive.

**STOP**

**END OF EXAM**

# Practice Test 1:
# Answers and Explanations

# PRACTICE TEST 1 ANSWER KEY

| | | | |
|---|---|---|---|
| 1. D | | 36. D | |
| 2. A | | 37. D | |
| 3. B | | 38. C | |
| 4. B | | 39. A | |
| 5. C | | 40. B | |
| 6. B | | 41. B | |
| 7. D | | 42. C | |
| 8. A | | 43. C | |
| 9. B | | 44. B | |
| 10. D | | 45. C | |
| 11. C | | 46. A | |
| 12. A | | 47. D | |
| 13. A | | 48. C | |
| 14. B | | 49. D | |
| 15. D | | 50. D | |
| 16. D | | 51. B | |
| 17. C | | 52. D | |
| 18. D | | 53. C | |
| 19. D | | 54. A | |
| 20. B | | 55. C | |
| 21. B | | 56. D | |
| 22. D | | 57. D | |
| 23. D | | 58. C | |
| 24. B | | 59. C | |
| 25. C | | 60. B | |
| 26. D | | 61. D | |
| 27. D | | 62. B | |
| 28. D | | 63. C | |
| 29. D | | 64. 425 | |
| 30. A | | 65. 1/4 or 0.25 | |
| 31. C | | 66. 240 | |
| 32. D | | 67. 0.3 | |
| 33. D | | 68. 1/2 | |
| 34. C | | 69. 525 | |
| 35. C | | | |

# PRACTICE TEST 1 EXPLANATIONS

## Section I: Multiple Choice

1. **D** The resting potential depends on active transport (the Na⁺K⁺-ATPase pump) and the selective permeability of the axon membrane to K⁺ than to Na⁺, which leads to a differential distribution of ions across the axonal membrane.

2. **A** The Krebs cycle occurs in the mitochondrial matrix. Don't forget to review the site of each stage of aerobic respiration. Glycolysis, the first step in aerobic respiration, occurs in the cytoplasm. The electron transport chain occurs along the inner mitochondrial membrane. Oxidative phosphorylation occurs as protons (H⁺ ions) move from the intermembrane space to the mitochondrial matrix.

3. **B** A heterotroph obtains its energy from organic molecules. An autotroph obtains energy from sunlight utilizing pigments such as chlorophyll, and uses $CO_2$ and water to make organic molecules. Therefore, (A) and (C) can be eliminated. Heterotrophs can obtain their energy from ingesting autotrophs, but they can also consume other heterotrophs. So you can eliminate (D), leaving (B) as your answer.

4. **B** In meiosis, the sister chromatids separate during the second metaphase of meiosis (Meiosis II), whereas the sister chromatids separate during metaphase of mitosis. Choice (A) is incorrect because in meiosis there are two rounds of cell division, whereas in mitosis there is only one round of cell division. Chromosomes are replicated during interphase in both meiosis and mitosis, so (C) is incorrect. Choice (D) can also be eliminated because spindle fibers form during prophase in both mitosis and meiosis.

5. **C** Amino acids are organic molecules that contain carbon, hydrogen, oxygen, and nitrogen, so eliminate (A), (B), and (D). Don't forget to associate amino acids with nitrogen because of the amino group $(NH_2)$.

6. **B** Unlike eukaryotes, prokaryotes (which include bacteria) do not contain membrane-bound organelles. Bacteria contain circular double-stranded DNA, ribosomes, and a cell wall, so (A) and (D) are incorrect. Also eliminate (C) because bacterial cell membranes are made up of a bilipid layer with proteins interspersed.

7. **D** Populations can be described as the evolutionary unit because changes in the genetic makeup of populations can be measured over time. Eliminate (A), as genetic changes occur only at the individual level. Only under Hardy-Weinberg equilibrium does the gene pool remain fixed over time in a population. However, this statement does not explain why the population is the evolving unit, so (B) is incorrect. Choice (C) is true but does not address the question.

8. **A** Positive feedback occurs when a stimulus causes an increased response. Choices (B), (C), and (D) are examples of negative feedback.

9.  **B**  In order to determine the genotype of the parents, use the ratio of the offspring given in the question and work backward. The ratio of black-haired to white-haired guinea pigs is 3:1. The offspring were therefore BB, Bb, Bb, and bb. Now draw a Punnett square to figure out the genotype of the parents. The parents are Bb and Bb.

10. **D**  The mean weight of the offspring in the next generation will be heavier than the mean weight of the original population because all the lighter horses in the original population died off. The normal distribution for weight will therefore shift to the heavier end (to the right of the graph). You can therefore eliminate (C) because the mean weight should increase. The mean weight of the offspring could be heavier or lighter than their parents, so you can also eliminate (A) and (B).

11. **C**  Apoptosis (programmed cell death), hox and homeotic genes (genes that control differentiation), and inductive effects (a tissue affecting the differentiation of another tissue) play a role in cell differentiation. Operons are sets of multiple genes regulated by a single regulatory unit in bacteria.

12. **A**  The primitive atmosphere lacked oxygen ($O_2$). It contained methane ($CH_4$), ammonia ($NH_3$), and hydrogen ($H_2$).

13. **A**  Villi and microvilli are fingerlike projections present in the small intestine that dramatically increase the surface area available for nutrient absorption.

14. **B**  Pepsin works in the stomach (not the small intestine) to break down proteins to peptides. Complete digestion occurs in the small intestine. Pancreatic lipase breaks down fats into three fatty acids and glycerol, so eliminate (A). Pancreatic amylase breaks down carbohydrates into simple sugars, so (C) can also be eliminated. Bile emulsifies lipids and makes them more accessible to lipase, so get rid of (D).

15. **D**  Ribosomes are the site of protein synthesis. Therefore, the correct answer should start with ribosome. So eliminate (A) and (C). The polypeptide then moves through the rough ER to the Golgi apparatus, where it is modified and packaged into a vesicle. The vesicle then floats to the plasma membrane and is secreted. Choice (D) is your answer.

16. **D**  This statement is false because an individual with two identical alleles is said to be *homozygous,* not heterozygous, with respect to that gene. Alleles are different forms of the same gene found on corresponding positions of homologous chromosomes, so (A) and (B) are incorrect. More than two alleles can exist for a gene, but a person can have only two alleles for each trait.

17. **C**  To make multiple copies of a plasmid (a small circular DNA), it should be inserted into a bacterium. A plasmid would not replicate if it were inserted into a virus, so eliminate (A). If a plasmid were treated with a restriction enzyme, it would be cut into smaller fragments. This would not give us cloned versions of the plasmid, so (B) is incorrect. If the plasmid were run on a gel (using gel electrophoresis), this would only tell us the size of the plasmid; therefore, (D) is also incorrect.

18. **D** The least likely explanation for why mutations are low is that mutations produce variability in a gene pool. Any gene is bound to mutate. This produces a constant input of new genetic information in a gene pool. This answer choice doesn't give us any additional information about the rate of mutations. Some mutations are subtle and cause only a slight decrease in reproductive output, so eliminate (A). Some mutations are harmful and decrease the productive success of the individual, so (B) is incorrect. Some mutations are deleterious and lead to total reproductive failure. The zygote fails to develop, Therefore, (C) can also be eliminated.

19. **D** The main challenge for organisms residing in hot arid environments is water loss. Long loops of Henle and storage of water in fatty tissues conserve water resources and large ears help disperse heat. Secretion is the elimination of water in the kidneys, which would not be advantageous for an organism in a dry environment.

20. **B** Legume plants are able to live in nitrogen-poor soil because they obtain nitrogen from nitrogen-fixing bacteria. These plants cannot make their own proteins without nitrogen from nitrogen-fixing bacteria, so eliminate (A), (C), and (D).

21. **B** Every cell in an organism has the same set of genes. Differences in the timing and expression levels of these genes leads to structural and functional differences.

22. **D** Deoxygenated blood from the vena cava enters the right atrium (2), then the right ventricle (1), and then enters the pulmonary arteries (4). The left atrium (5), left ventricle (6), and aorta (3) all carry oxygenated blood.

23. **D** The egg yolk provides food for the embryo. The amnion protects the embryo, so (A) is incorrect. The chorion is the outermost layer that surrounds the embryo, so eliminate (B). Choice (C) can also be eliminated because the placenta provides nourishment for the embryo in mammals, not in birds.

24. **B** Normal cells can become cancerous when a virus invades the cell and takes over the replicative machinery. A pilus forms between two bacteria, so (A) is wrong. Also eliminate (C) because the host's genome is not converted to the viral genome. Choice (D) is incorrect because spores are released by fungi, not viruses.

25. **C** The primary difference between bone and cartilage is that cartilage is flexible and lacks blood vessels. They are both part of the skeletal system; get rid of (A). Bone and cartilage are connective tissues made up of collagen and calcium salts, so eliminate (B). Both cartilage and bones secrete a matrix, making (D) incorrect as well.

26. **D** If potential mates have similar breeding seasons they will most likely mate. Use common sense to eliminate the other answer choices. If the organisms don't meet, they won't reproduce; eliminate (A). Also eliminate (B) and (C): If potential mates do not share the same behaviors (such as courtship rituals), they may not mate.

27. **D** There is no union of gametes in mitosis. Choices (A) and (C) are incorrect: Asexual reproduction involves the production of two new cells with the same number of chromosomes as the parent cell. If the parent cell is diploid, then the daughter cells will be diploid. The daughter cells are identical to the parent cell, so eliminate (B).

28. **D** Arteries are thick-walled vessels that carry blood *away* from the heart. Blood moves by contracting muscles. Choices (A), (B), and (C) are all characteristics of veins. Veins are thin-walled vessels (with valves) that return blood to the heart.

29. **D** Cohesion, adhesion, and capillary action are special properties of water that are necessary for transpiration to occur. Water is a hydrophilic, polar molecule.

30. **A** Crossing-over (exchange of genetic material) occurs in prophase I of meiosis. Eliminate (B) because during metaphase I, the tetrads line up at the metaphase plate. Choice (C) is also incorrect; during anaphase, the tetrads separate. Finally, you can eliminate (D) because during prophase II, the chromosomes become visible again.

31. **C** Crossing-over and synapsis occur during meiosis, which produces haploid gametes. Separation of homologous chromosomes occurs during meiosis I, while separation of sister chromatids does not occur until meiosis II.

32. **D** Tropism describes the reaction of plants to a stimulus. Gravitropism specifies a reaction to gravity, and negative specifies that the reaction is away from the stimulus.

33. **D** The net primary productivity is what the energy producers have left for storage after their own energy needs have been met. Energy stored = (Total energy produced) – (energy used in cellular respiration). Choice (A) is incorrect; it is not the energy available to producers. You can also eliminate (B) because it is not the total chemical energy in producers—it's the gross primary productivity. Get rid of (C) as well because it is not the biomass (total living material) among producers.

34. **C** Choice (A) and (B) are incorrect: These organisms exhibit the same behavior because they were subjected to the same environmental conditions and similar habitats. This is an example of convergent evolution. However, they are not genetically similar, so (C) is the answer. (One is an insect and the other a bird.) They are analogous, so (D) is also incorrect. They exhibit the same function but are structurally different.

35. **C** A symbiotic relationship is a relationship among two organisms in which both benefit. Choice (A) describes parasitism, and (B) describes commensalism. Choice (D) is an example of altruistic behavior.

36. **D** Beta cells secrete insulin. Destruction of beta cells in the pancreas will halt the production of insulin. Therefore, eliminate (A), (B), and (C). This will lead to an increase in blood glucose levels.

37. **D** The sympathetic division is active during emergency situations. This leads to a decrease in peristalsis in your gastrointestinal tract. (Your stomach shuts down.) Stimulation of the sympathetic nervous system leads to (A), pupils dilating, (B), peripheral blood vessels constricting, and (C), sweating.

38. **C** This is an example of mutualism. Both organisms benefit. Choice (A), parasitism, is an example of a symbiotic relationship in which one organism benefits and the other is harmed. Choice (B), commensalism, is when one organism benefits and the other is unaffected. Choice (D), endosymbiosis, is the idea that some organelles originated as symbiotic prokaryotes that live inside larger cells.

39. **A** They both contain genetic material (DNA), a plasma membrane, and a cell wall. Unlike fungi, bacteria lack a definite nucleus. Therefore, eliminate (B). Bacteria are unicellular, whereas fungi are both unicellular and multicellular. Therefore, eliminate (C) and (D).

40. **B** The two hormones that are responsible for the maintenance of $Ca^{2+}$ in the blood are parathyroid hormone and calcitonin. Therefore, eliminate (A) and (C). Now review the effects of these hormones. Parathyroid hormone increases $Ca^{2+}$ in the blood, while calcitonin decreases $Ca^{2+}$ in the blood. A sustained decrease in circulating $Ca^{2+}$ levels might be caused by decreased levels of parathyroid hormone.

41. **B** This question tests your understanding of what stage is responsible for the synthesis of new proteins for lactose utilization. The region of bacterial DNA that controls gene expression is the *lac* operon. Structural genes will be transcribed to produce enzymes, which produce an mRNA involved in digesting lactose.

42. **C** Trypsin is an enzyme, which are proteins and organic catalysts that speed up reactions without altering them. They are not consumed in the process. Therefore, you can eliminate (A) and (B). The rate of reaction can be affected by the concentration of the substrate up to a point, so (D) can be eliminated.

43. **C** DNA polymerase, not RNA polymerase, is the enzyme that causes the DNA strands to elongate. DNA helicase unwinds the double helix, so (A) is true and therefore incorrect. Choice (B), DNA ligase seals the discontinuous Okazaki fragments, is also true. Eliminate it. In the presence of DNA polymerase, DNA strands always grow in the 5' to 3' direction as complementary bases attach. Therefore, (D) is also incorrect.

44. **B** A voltage change from +50 to –70 is called repolarization. You can eliminate (A) because a voltage change from –70 to +50 is called depolarization. Choice (C) is incorrect; a voltage change from –70 to –90 is called hyperpolarization. Choice (D) can be eliminated as well; an action potential is a traveling depolarized wave. It refers to the whole thing, from depolarization, repolarization, hyperpolarization, and back to a resting potential.

45. **C** Electrons passed down along the electron transport chain from one carrier to another lose energy and provide energy for making ATP. Glucose is decomposed during glycolysis, but this process is not associated with energy given up by electrons; eliminate (A). Glucose is made during photosynthesis, so eliminate (B). NADH is are energy-rich molecule, which accepts electrons during the Krebs cycle. Therefore, (D) is incorrect as well.

46.  **A**  The oxygen released during the light reaction comes from the splitting of water. (Review the reaction for photosynthesis.) Therefore, water must have originally contained the radioactive oxygen. Carbon dioxide is involved in the dark reaction and produces glucose, so eliminate (B). Glucose is the final product and would not be radioactive unless carbon dioxide was the radioactive material, so (C) is incorrect. Finally, eliminate (D), because nitrogen is not part of photosynthesis.

47.  **D**  Hydrophobic means "water fearing," so water cannot be a hydrophobic solvent. Water is a polar solvent; it contains both negatively and positively charged ends that can form hydrogen bonds with other polar substances. Water can break up into $H^+$ and $OH^-$ ions.

48.  **C**  Pheromones act as sex attractants, alarm signals, or territorial markers. Auxins are plant hormones that promote growth, so eliminate (A). Hormones are chemical messengers that produce a specific effect on target cells within the same organism, so eliminate (B). Enzymes are catalysts that speed up reactions, so eliminate (D).

49.  **D**  Homologous structures are organisms with the same structure but different functions. The forelegs of an insect and the forelimbs of a dog are not structurally similar. (One is an invertebrate and the other a vertebrate.) They do not share a common ancestor. However, both structures are used for movement. All of the other examples are vertebrates that are structurally similar.

50.  **D**  Muscle contractions require calcium ions, so (I) is required. You need energy order to have a muscle contraction, so (II) is required. And actin is one of the proteins involved in muscle contractions, so (III) is required. The answer is (D).

51.  **B**  Speciation occurred in the Galapagos finches as a result of the different environments on the islands. This is an example of divergent evolution. The finches were geographically isolated. Choice (A), convergent evolution, is the evolution of similar structures in distantly related organisms. Choice (C), disruptive selection, is selection that favors both extremes at the expense of the intermediates in a population. Choice (D), stabilizing selection, is selection that favors the intermediates at the expense of the extreme phenotypes in a population.

52.  **D**  Mutations produce genetic variability. All of the other answer choices are forms of asexual reproduction.

53.  **C**  Secondary consumers feed on primary consumers. If you set up a pyramid of numbers, you'll see that the herrings belong to the third trophic level.

54.  **A**  The biomass is the total bulk of a particular living organism. The plankton population has both the largest biomass and the most energy.

55.  **C**  If the herring population decreases, this will lead to an increase in the number of crustaceans and a decrease in the plankton population. Reorder the organisms according to their trophic levels and determine which populations will increase and decrease accordingly.

56.  **D**  This question tests your ability to trace the neural pathway of a motor (effector) neuron. The nerve conduction will travel from the spinal cord (where interneurons are located) to the muscle.

57. **D** Because the brain is destroyed, it is not associated with the movement of the leg. Choice (A) is incorrect, as reflex actions are automatic. Choices (B) and (C) can also be eliminated; both of these statements are true but are not supported by the experiment.

58. **C** Neurotransmitters are released from the axonal bulb of one neuron and diffuse across a synapse to activate a second neuron. The second neuron is called a postsynaptic neuron. A neurotransmitter can either excite or inhibit the postsynaptic neuron. The myelin sheath speeds up the conduction in a neuron, so (A) is wrong. Also eliminate (B) and (D), as both sodium and potassium channels open during an action potential. Neurotransmitters are not involved in actions related to the axon membrane. They do not force potassium ions to move against a concentration gradient.

59. **C** The DNA template strand is complementary to the mRNA strand. Using the mRNA strand, work backward to establish the sequence of the DNA strand. Don't forget that DNA strands do not contain uracil, so eliminate (B) and (D).

60. **B** Use the amino acid chart to determine the sequence after uracil is deleted. The deletion of uracil creates a frameshift.

61. **D** The mRNA is modified before it leaves the nucleus. It becomes smaller when introns (intervening sequences) are removed. A poly(A) tail and a cap are added to the mRNA and would therefore increase the length of the mRNA, so you can eliminate (A) and (B). Choice (C) is also incorrect, as exons are the coding sequences that are kept by the mRNA.

62. **B** The Weddell seal stores twice as much oxygen as humans. Calculate the liters per kilograms weight for both the seal and man using the information at the bottom of the chart. The Weddell seal stores 0.058 liters/kilograms (25.9 liters/450 kilograms) compared to 0.028 liters/kilograms (1.95 liters/70 kilograms) in humans.

63. **C** The most plausible answer is that blood is redirected toward the central nervous system, which permits the seal to navigate for long durations. Choice (A) is incorrect; the seal does not need to increase the number of red blood cells in the nervous system. Choice (B) can also be eliminated, as the seal does not need to increase the amount of oxygen to the skeletal system. Eliminate (D) because the diversion of blood does not increase the concentration of oxygen in the lungs.

64. **425** If you look at the absorption spectrum, you'll see that chlorophyll a has two peaks, one at 425 nm and one at 680 nm. Chlorophyll *a* maximally absorbs light at approximately 425 nm.

65. $\frac{1}{4}$ or **0.25**

Make a Punnett square to determine the probability that the couple has a child with blood type AB. The probability is $\frac{1}{4}$ whether it's the first child or the third child.

66. **240** 24,000 lbs of elk is equivalent to 12 tons since there are 2,000 lbs per ton. With a 5% efficiency of transfer from their food source, that 12 tons represents 5% of the weight of the plants they would need to consume. 5/100 = 12/tons needed, making the answer 240.

67. **0.3** The frequency of the homozygous dominant genotype (AA) is 0.49. To find the dominant *allele* frequency, we can use the formula provided by the Hardy-Weinberg theory, $p^2 + 2pq + p^2 = 1$, where $p$ represents the dominant allele and $q$ represents the recessive allele. Because we know that $p^2$ represents the frequency of the homozygous dominant genotype, we can find the frequency of the dominant allele ($p$) by taking the square root of the frequency of the genotype. The square root of 0.49 is 0.7. Note that by using the formula $p + q = 1$, we can also determine the frequency of the recessive allele ($q$). It would be 0.3 ($0.7 + q = 1$).

68. $\dfrac{1}{2}$ Draw a Punnett square for the couple to determine the probability of color-blindness for the boys.

Individuals 6 and 7, ($X^cX$ and XY) will produce two males: one $X^cY$ and one XY. The probability of color-blindness is therefore $\dfrac{1}{2}$.

69. **525** Make sure you read the question carefully. You are asked to calculate the number of liters per week, not per day. The chart tells us that a coconut palm loses 75 liters a day, which would mean 525 liters a week ($7 \times 75 = 525$).

# Section II: Free Response

On the following pages you'll find two types of aids for grading your essays: checklists and sample responses. Checklists like these are used to grade your essays on the actual test. They're actually quite simple to use.

For each item you mentioned, give yourself the appropriate number of points (1 point, 2 points, and so on). Remember that you can only get a maximum of 10 points for each essay. As you evaluate your work, don't be kind to yourself just because you like your own essay. If the checklist mentions an explanation of structure and function and you failed to give both, then do not give yourself a point. ETS is very particular about this. You need to mention precisely the things they've listed, in the way they've listed them, in order to gain points. What if something you've mentioned doesn't appear on the list?

Provided you know it's valid, give yourself a point. If it was something you pulled out of your hat at the last minute, odds are it's not directly applicable to the question. However, because you might come up with details even more specific than those contained in the checklist, it's not unlikely that the example or structure you've cited is perfectly valid. If so, go ahead and give yourself the point. Remember that ETS hires college professors and high school teachers to read your essays, so they'll undoubtedly recognize any legitimate information you slip into the essay.

The second part of the answer key involves short paragraphs. These are not templates, ETS does not expect you to write this way. They are simply additional tools to help illustrate the things you need to squeeze into your essay in order to rack up the points. They explain in some detail how the various parts of the checklist relate to one another, and may give you an idea about how best to integrate them into your own essays come test-time.

If you find that you didn't do too well on the essay portion, be sure to carefully read the chapter on the free-response questions (Chapter 14). You can use the chapters themselves as checklists. If you find it too difficult to grade your own essays, see if your teacher or a classmate will help you out. Good luck!

# Long-Form Free Response Checklist 1

a. **Cell size ratio**— 4 points maximum

   **(2 points each)**

   A higher ratio of surface area to volume allows for greater space for solutes to move in an out of cells.

   Cells must maintain homeostasis and in order to do this must eliminate wastes, ingest nutrients, and maintain osmotic and ion balances.

   As a cell grows larger, this ratio diminishes, limiting cell size.

**b.  Adaptations**—4 points maximum

   **(2 points each)**
   Alveoli
   Convoluted membranes in chloroplasts and mitochondria
   Root hairs
   Villi and microvilli
   Endoplasmic reticulum

**c.  Transport across the membrane**

   **(2 points each)**
   Small polar molecules cross by facilitated diffusion and require membrane channels.
   Water is an example of a small polar molecule that crosses through channels
      called aquaporins.
   Ions are polar species that cross the cell membrane via protein channels.
   Small nonpolar molecules can freely diffuse across the lipid bilayer.
   Seed/seedcoat
   Examples of small nonpolar molecules include $CO_2$, $O_2$, and steroid hormones.

# Sample Long-Form Free Response 1

Cells need a certain surface area to volume ratio to exchange materials with the environment in order to maintain homeostasis. A high surface area to volume ratio allows cells to regulate ion concentrations, as well as take in nutrients and eliminate wastes. As a cell becomes larger, the ratio of surface area to volume decreases, and if a cell grows too large it cannot carry out these functions and therefore will not survive. Organisms have adapted a variety of ways to increase surface area. One example is alveoli, small air sacs in the lungs that maximize surface area in order to allow for maximum respiration. Another example of increasing surface area is villi in the small intestine, which results in a large surface area for the absorption of nutrients.

   Depending on polarity, molecules cross the cell membrane in different ways. Small nonpolar molecules, like carbon dioxide, can freely diffuse through the cell membrane. However, due to the hydrophobic nature of the interior of the cell membrane, small polar molecules need protein channels to allow selective diffusion into or out of the cell. One example of a polar molecule is water, which crosses the membrane through channels called aquaporins.

# Long-Form Free Response Checklist 2

a. **Type of Mutation**—4 points maximum

   **(1 point each)**

   Base substitution (involves one nucleotide being replaced by another)

   Change in DNA/codon of mRNA draws new tRNA molecule

   New tRNA carries wrong amino acid, altering polypeptide

   Change in polypeptide, in turn, alters hemoglobin protein

   Distorted red blood cell cannot efficiently carry oxygen

b. **Gel Electrophoresis**—5 points maximum

   **How it works**—3 points maximum
   **(1 point each)**

   | | |
   |---|---|
   | Apparatus | DNA is put into wells of an agarose gel with a buffer |
   | Electricity | Electrical potential (electrical charge moves fragments) |
   | Charge | Negatively charged fragments move toward the positive pole |
   | Size/Molecular weight | Smaller fragments move faster across the gel |

   **How it can be applied to identify the mutant gene**—2 points maximum
   **(1 point each)**

   Compare the normal hemoglobin to the abnormal hemoglobin

   Use restriction enzymes to cut the DNA into several fragments

   Normal hemoglobin can be used as a marker

   Samples of the two proteins, or DNA, should be identical except for the source in order to compare normal and mutant forms.

c. **Punnett Square**—2 points maximum

   Sickle-cell disease is an autosomal recessive disorder. Gender does not matter.

   A normal female could be heterozygous or homozygous dominant.

   A male with the disease would be homozygous.

## Sample Long-Form Free Response 2

Sickle cell anemia is a disease in which the red blood cells have an abnormal shape because of a base substitution. Base substitution involves an error in DNA replication in which one nucleotide is replaced by another nucleotide. This causes the codon of an mRNA to contain an incorrect base. The codon, therefore, matches up with the anticodon of a different tRNA. This tRNA carries a different amino acid. The change in the amino acid alters the polypeptide. The polypeptide, in turn, alters the hemoglobin protein. In this case, the distorted red blood cell cannot efficiently carry oxygen.

Biologists must have determined the nature of the hemoglobin mutation by comparing the normal hemoglobin gene to the abnormal hemoglobin gene, using the technique gel electrophoresis. Gel electrophoresis identifies the difference between two molecules by examining the different rates each molecule moves across the gel. Substances move across a gel according to their molecular weight. For example, smaller fragments move faster than larger fragments. The biologists must have placed fragments of the two genes on an agarose gel and used the normal hemoglobin gene as the marker—the source of comparison to the abnormal hemoglobin gene. (A restriction enzyme was used to cut the two DNA sequences into several fragments prior to loading the gel.) The two DNA sequences should have been identical except for the fragment that contained the abnormal gene.

If a normal noncarrier female mates with a male who is homozygous for the disease, these are the results using a Punnett square,

All of the offspring would be heterozygotes. For sickle-cell anemia, these offspring would be carriers.

## Short-Form Free Response Checklist 3

a. **Ions associated with plasma membrane**—4 points maximum

**(1 point each)**

| Ions | Description |
|------|-------------|
| Selectively permeable to K$^+$ ions | (Concentrated on the inside of the cell) |
| Impermeable to Na$^+$ ions | (Concentrated on the outside of the cell) |
| Build up of positively charged ions outside of the membrane | (Inside is negatively charged) |
| Sodium-potassium pump | (Maintains the ion gradient) |

b. **Role of the membrane**—4 points maximum

**(.5 point each)**

Nerve impulses are electrochemical (associated with ion and electrical changes in plasma membrane)

Action potential (change in the membrane potential)

Resting stage (voltage charge is –70 millivolts)

Depolarization (Na$^+$ moves into the cell down its cell gradient)

Repolarization (K$^+$ ions move out of the cell down their cell gradient)

Sodium channels (membrane channel that is voltage-gated)

Potassium channels (membrane channel that is voltage-gated)

Sodium-potassium pump (membrane protein that maintains ion gradient)

## Sample Short-Form Free Response 3

A nerve cell is fundamentally similar in structure and function to other somatic cells. Like other cells, a nerve cell consists of a bilipid layer made up of phospholipids and proteins. This membrane is semipermeable; it regulates the passage of certain ions across its membrane. The cell body of a nerve cell also contains cellular structures such as a nucleus, which regulates the activities of the cell, and ribosomes, which make proteins.

The plasma membrane of a resting neuron is selectively permeable to $K^+$ ions and impermeable to $Na^+$ ions. Because of this selective permeability, the $K^+$ ions are concentrated on the inside of the cell and are able to slowly diffuse outward, while the $Na^+$ ions are concentrated on the outside of the cell. This leads to a build up of positively-charged ions outside of the neuronal membrane. The inside of the plasma membrane is therefore negatively charged. These chemical gradients are maintained by a sodium-potassium pump. This pump actively moves three $Na^+$ ions out of the cell for every two $K^+$ ions brought into the cell.

The neuronal membrane plays an important role in the conduction of a nerve impulse. Nerve impulses are electrochemical. This means, the forces that cause ions to move across a membrane are both a concentration gradient and an electrical gradient. The membrane is associated with two chemicals ($Na^+$ ions and $K^+$ ions) as well as a change in the voltage charge. When a nerve cell is undisturbed, the membrane is said to be in a resting stage. The membrane is polarized and the voltage charge is −70 millivolts. During a nerve impulse, there is a change in the membrane permeability, $Na^+$ ions rush into the cell, and the inside becomes more positively charged. The membrane is now said to be depolarized. $Na^+$ move into the cell down its concentration gradient. Next, the $Na^+$ channels close and $K^+$ channels open and $K^+$ ions move out of the cell. The cell is now said to be repolarized. The original ion concentration is reestablished by the sodium-potassium pump.

## Short-Form Free Response Checklist 4

a. **Krebs cycle, the electron transport chain, oxidative phosphorylation as aerobic processes**—2 points maximum

> They require oxygen
>
> They cannot occur under anaerobic conditions

b. **Site of each step**—3 points maximum

| Stage | Site |
|---|---|
| Krebs cycle | Mitochondrial matrix |
| Electron transport chain | Along the inner mitochondrial membrane |
| Oxidative phosphorylation | As hydrogens move from the intermembrane space to the mitochondrial matrix |

## Sample Short-Form Free Response 4

a. The Krebs cycle, electron transport chain, and oxidative phosphorylation are all part of aerobic respiration. During aerobic respiration, glucose is completely converted to $CO_2$, ATP, and water. These steps are considered aerobic processes because they cannot occur under anaerobic conditions; they require oxygen. Glycolysis, on the other hand, can occur under both aerobic and anaerobic conditions.

b. These three stages of aerobic respiration occur in different parts of the mitochondria. The Krebs cycle occurs in the mitochondrial matrix. Two acetyl CoA enter the Krebs cycle and produce NADH, $FADH_2$, ATP, and $CO_2$. The products of the Krebs cycle (NADH and $FADH_2$) are sent to the electron transport chain. The electron transport chain occurs along the inner mitochondrial membrane. The final stage of aerobic respiration—oxidative phosphorylation—occurs as hydrogens move across from the intermembrane space to the mitochondrial matrix.

## Short-Form Free Response Checklist 5

**(1 point each)**

Produces somatic cells versus produces gametes

One round of division versus two rounds of division

Meiosis only: recombination between homologous chromosomes


## Sample Short-Form Free Response 5

The difference between meiosis and mitosis arises because of the different goals of these cellular processes. First, mitosis is designed to reproduce copies of somatic cells, while meiosis is designed to create gametes with greater variability. Second, in order to create the appropriate level of genetic information, meiosis requires two rounds of division whereas recreating somatic cells only requires one round. Third, in order to increase genetic variability, recombination can occur as part of meiosis between homologous chromosomes. Such mixing is not part of mitosis since the goal is to simply recreate the original cell.


## Short-Form Free Response Checklist 6

**(1 point each)**

Definition of homologous structures

Relevant example (many possible)


## Sample Short-Form Free Response 6

Homologous structures are those with similar structures because they arose from the same ancestral source. An example of homologous structures would be the arm of a human and the front leg of a cat. Their functions are different in their current form, but the ancestral origin is shared.

## Short-Form Free Response Checklist 7

**(1 point each)**

Transduction

Transformation

**(2 points each)**

Conjugation

## Sample Short-Form Free Response 7

There are three ways in which bacteria transmit genetic information laterally to introduce new phenotypes. These are transduction, transformation, and conjugation. Transduction is the transmission of genetic material from one bacteria to another via a lysogenic virus. Transformation is the uptake of naked DNA from the environment. Conjugation is the cell-to-cell transfer of genetic material in the form of plasmids across a pilus formed between two cells.

## Short-Form Free Response Checklist 8

**(1 point each)**

No respiration

Not able to create own energy

Not able to reproduce independently

Not able to live outside host cells

## Sample Short-Form Free Response 8

Viruses are typically not considered to be alive because they are not able to engage in many of the functions associated with life. Though viruses must engage in processes like translation that require energy, they do not engage in any process to create that energy. In that same line, there is no respiration of any kind being done by viruses. Further, viruses have no existence outside of their host cells; no viral functions happen outside of the context of a cell. Due to that restriction, viruses are also not able to reproduce themselves. Being entirely reliant on a host cellular environment for reproduction is another reason to consider them as not alive.

# Part III
# About the AP Biology Exam

- The Structure of the AP Biology Exam
- How the AP Biology Exam Is Scored
- Overview of Content Topics
- How AP Exams Are Used
- Other Resources
- Designing Your Study Plan

# THE STRUCTURE OF THE AP BIOLOGY EXAM

The AP Biology Exam is three hours long and is divided into two sections: Section I (multiple-choice questions) and Section II (free-response questions).

Section I consists of 69 questions. These are broken down into Part A (63 multiple-choice questions) and Part B (6 grid-in questions). The multiple-choice questions are further broken down into two parts: regular multiple-choice questions and questions dealing with experiments or data.

Section II involves free-response questions. You'll be presented with two long-form free-response questions and six short-form free-response questions touching upon key issues in biology. You'll be given a 10-minute reading period followed by 80 minutes to answer all eight questions.

If you're thinking that this sounds like a heap of work to try to finish in three hours, you're absolutely right. How can you possibly tackle so much science in so little time? Fortunately, there's absolutely no need to. As you'll soon see, we're going to ask you to leave a small chunk of the test blank. Which part? The parts you don't like. This selective approach to the test, which we call "pacing," is probably the most important part of our overall strategy. But before we talk strategy, let's look at the topics that are covered by the AP Biology Exam.

# HOW THE AP BIOLOGY EXAM IS SCORED

AP scores are calculated from your scores on the multiple-choice and free-response sections. The final score is reported on a scale from 1 to 5. The following table explains what that final score means:

| Score (Meaning) | Number of test takers receiving this score* | Percentage of test takers receiving this score* | Equivalent grade in a first-year college course | Credit granted for this score? |
|---|---|---|---|---|
| 5 (extremely qualified) | 14,099 | 7% | A | Most schools grant credit. |
| 4 (well qualified) | 47,765 | 22% | A–, B+ | Most schools grant credit. |
| 3 (qualified) | 75,010 | 35% | B–, C | Some schools grant credit, some do not. |
| 2 (possibly qualified) | 57,815 | 27% | C, D | Very few schools grant credit. |
| 1 (no recommendation) | 18,605 | 9% | D | No |

*The data above is from the College Board website and based on the May 2014 test administration.

Remember that colleges' rules may vary when it comes to granting credit for AP courses. You should contact the individual admissions departments to find out what score you need on the exam to ensure you'll be given credit.

For the AP Biology Exam, your scores on the multiple-choice section and free-response section are each worth 50 percent of your final score. On the multiple-choice section, your total score is based on the number of questions answered correctly, and you do not lose any points for incorrect answers. Unanswered questions do not receive points. The free-response section is graded on separate point system; long-form responses (Questions 1 and 2) are graded on a 9-point scale, while short-form responses can received a maximum of 4 points. Your scores are tallied to determine your total free-response score. The free-response score is then combined with your multiple-choice score and weighted to figure out where your score falls within the standard AP scoring scale (1 to 5).

# OVERVIEW OF CONTENT TOPICS

The AP Biology Exam covers these four Big Ideas:

- **Big Idea 1:** The process of evolution drives the diversity and unity of life.

- **Big Idea 2:** Biological systems utilize free energy and molecular building blocks to grow, to reproduce, and to maintain dynamic homeostasis.

- **Big Idea 3:** Living systems store, retrieve, transmit, and respond to information essential to life processes.

- **Big Idea 4:** Biological systems interact, and these systems and their interactions possess complex properties.

These four areas are further subdivided into major topics. These topics include the following:

- Chemistry of Life
  - Organic molecules in organisms
  - Water
  - Free-energy changes
  - Enzymes
- Cells
  - Prokaryotic and eukaryotic cells
  - Membranes
  - Subcellular organization
  - Cell cycle and its regulation

- Cellular Energetics
  - Coupled reactions
  - Fermentation and cellular respiration
  - Photosynthesis
- Heredity
  - Meiosis and gametogenesis
  - Eukaryotic chromosomes
  - Inheritance patterns
- Molecular Genetics
  - RNA and DNA structure and function
  - Gene regulation
  - Mutation
  - Viral structure and replication
  - Nucleic acid technology and applications
- Evolutionary Biology
  - Early evolution of life
  - Evidence for evolution
  - Mechanism of evolution
- Diversity of Organisms
  - Evolutionary patterns
  - Survey of the diversity of life
  - Phylogenetic classification
  - Evolutionary relationships
- Structure and Function of Plants and Animals
  - Reproduction, growth, and development
  - Structural, physiological, and behavioral adaptation
  - Response to the environment
- Ecology
  - Population dynamics
  - Communities and ecosystems
  - Global issues

This might seem like an awful lot of information. But for each topic, there are just a few key facts you'll need to know. Your biology textbooks may go into far greater detail about some of these topics than we do. That's because they're trying to teach you "correct science," whereas we're aiming to improve your scores. Our science is perfectly sound; it's just cut down to size. We've focused on crucial details and given you only what's important. Moreover, as you'll soon see, our treatment of these topics is far easier to handle.

## HOW AP EXAMS ARE USED

Different colleges use AP Exams in different ways, so it is important that you go to a particular college's website to determine how it uses AP Exams. The three items below represent the main ways in which AP Exam scores can be used:

- **College Credit.** Some colleges will give you college credit if you score well on an AP Exam. These credits count toward your graduation requirements, meaning that you can take fewer courses while in college. Given the cost of college, this could be quite a benefit, indeed.

- **Satisfy Requirements.** Some colleges will allow you to "place out" of certain requirements if you do well on an AP Exam, even if they do not give you actual college credits. For example, you might not need to take an introductory-level course, or perhaps you might not need to take a class in a certain discipline at all.

- **Admissions Plus.** Even if your AP Exam will not result in college credit or allow you to place out of certain courses, most colleges will respect your decision to push yourself by taking an AP Course or an AP Exam outside of a course. A high score on an AP Exam shows mastery of more difficult content than is taught in many high school courses, and colleges may take that into account during the admissions process.

## OTHER RESOURCES

There are many resources available to help you improve your score on the AP Biology Exam, not the least of which are your **teachers**. If you are taking an AP class, you may be able to get extra attention from your teacher, such as obtaining feedback on your essays. If you are not in an AP course, reach out to a teacher who teaches AP Biology and ask if the teacher will review your essays or otherwise help you with content.

**Bring on the Bio!**
If you want extra practice, remember to check out our supplemental title, *550 AP Biology Practice Questions*.

Another wonderful resource is **AP Central**, the official site of the AP Exams. The scope of the information at this site is quite broad and includes:

- a course description, which includes details on what content is covered and sample questions
- sample questions from the AP Biology Exam
- free-response question prompts and multiple-choice questions from previous years

The AP Central home page address is: **http://apcentral.collegeboard.com/.**

For up-to-date information about the ongoing changes to the AP Biology Exam Course, please visit: **http://apcentral.collegeboard.com/apc/public/courses/teachers_corner/2117.html**

Finally, The Princeton Review offers tutoring and small group instruction. Our expert instructors can help you refine your strategic approach and add to your content knowledge. For more information, call 1-800-2REVIEW.

**More Test Prep and Practice**
Head over to PrincetonReview.com for a host of test prep resources, including PSAT, SAT, and ACT practice!

# DESIGNING YOUR STUDY PLAN

As part of the Introduction, you identified some areas of potential improvement. Let's now delve further into your performance on Practice Test 1, with the goal of developing a study plan appropriate to your needs and time commitment.

Read the answers and explanations associated with the multiple-choice questions (starting at page 29). After you have done so, respond to the following questions:

- Review the bulleted list of topics on pages 49 and 50. Next to each topic, indicate your rank of the topic as follows: "1" means "I need a lot of work on this," "2" means "I need to beef up my knowledge," and "3" means "I know this topic well."

- How many days/weeks/months away is your exam?

- What time of day is your best, most focused study time?

- How much time per day/week/month will you devote to preparing for your exam?

- When will you do this preparation? (Be as specific as possible: Mondays and Wednesdays from 3:00 to 4:00 P.M., for example)

- Based on the answers above, will you focus on strategy (Part IV) or content (Part V) or both?

- What are your overall goals in using this book?

# Part IV
# Test-Taking Strategies for the AP Biology Exam

- Preview
1 How to Approach Multiple-Choice Questions
2 How to Approach Free-Response Questions
3 Using Time Effectively to Maximize Points
- Reflect

## PREVIEW

Review your responses to the first three questions on page 2 and then respond to the following questions:

- How many multiple-choice questions did you miss even though you knew the answer?

- On how many multiple-choice questions did you guess blindly?

- How many multiple-choice questions did you miss after eliminating some answers and guessing based on the remaining answers?

- Did you find any of the free-response questions easier/harder than the others—and, if so, why?

## HOW TO USE THE CHAPTERS IN THIS PART

For the following Strategy chapters, think about what you are doing now before you read the chapters. As you read and engage in the directed practice, be sure to appreciate the ways you can change your approach. At the end of Part IV, you will have the opportunity to reflect on how you will change your approach.

# Chapter 1
# How to Approach
# Multiple-Choice
# Questions

# SECTION I

As we mentioned earlier, the multiple-choice section consists of the following two parts:

- Part A—63 standard multiple-choice questions and multiple-choice questions dealing with an experiment or a set of data

- Part B—6 grid-in questions

# Part A

Part A of Section I consists of 63 run-of-the-mill multiple-choice questions. These questions test your grasp of the fundamentals of biology and your ability to apply biological concepts to help problem-solve. Here's an example:

> If a segment of DNA reads 5'-ATG-CCA-GCT-3', the mRNA strand that results from the transcription of this segment will be
>
> (A)    3'-TAC-GGT-CGA-5'
>
> (B)    3'-UAC-AGT-CAA-5'
>
> (C)    3'-TAA-GGU-CGA-5'
>
> (D)    3'-TAC-GGT-CTA-5'

Don't worry about the answer to this question for now. By the end of this book, it will be a piece of cake. The majority of the questions in this section are presented in this format. A few questions may include a figure, a diagram, or a chart.

The second part of the first portion also consists of multiple-choice questions, but here you're asked to think logically about different biological experiments or data.

Here's a typical example:

Questions 60 and 61 refer to the following diagram and information.

To understand the workings of neurons, an experiment was conducted to study the neural pathway of a reflex arc in frogs. A diagram of a reflex arc is given below.

Neural Action

60. Which of the following represents the correct pathway taken by a nerve impulse as it travels from the spinal cord to effector cells?

(A)    1-2-3-4

(B)    6-5-4-3

(C)    2-3-4-5

(D)    4-5-6-7

61. The brain of the frog is destroyed. A piece of acid-soaked paper is applied to the frog's skin. Every time the piece of paper is placed on its skin, one leg moves upward. Which of the following conclusions is best supported by the experiment?

(A)    Reflex actions are not automatic.

(B)    Some reflex actions can be inhibited or facilitated.

(C)    All behaviors in frogs are primarily reflex responses.

(D)    This reflex action bypasses the central nervous system.

You'll notice that these particular questions refer to an experiment. Many of the questions in this portion test your ability to integrate information, interpret data, and draw conclusions from the results.

## Part B

Part B of Section I consists of six grid-in questions in which an answer needs to be calculated based on information presented in the question. That numeric response is then filled in on a grid and bubbled accordingly. Answers can be in the form of integers, decimals, or fractions. A four-function calculator can be used on these questions. Here's a typical example:

If the genotype frequencies of an insect population are AA = 0.49, Aa = 0.42, and aa = 0.09, what is the gene frequency of the dominant allele?

We mentioned earlier that our approach is strategy-based. As you're about to see, many of these strategies are based on common sense—for example, using mnemonics like "ROY G. BIV." (Remember that one? It's the mnemonic for red, orange, yellow, green, blue, indigo, violet—the colors of the spectrum.) Others do not make as much sense. In fact, we're going to ask you to throw out much of what you've been taught when it comes to taking standardized tests.

## PACE YOURSELF

When you take a test in school, how many questions do you answer? Naturally, you try to answer all of them. You do this for two reasons: (1) Your teacher told you to, and (2) if you left a question blank, your teacher would mark it wrong. However, that's not the case when it comes to the AP Biology Exam. In fact, finishing the test is the worst thing you can do. Before we explain why, let's talk about timing.

One of the main reasons that taking the AP Biology Exam can be so stressful is the time constraint we discussed above—75 seconds per multiple-choice question and 21 minutes per essay. If you had all day, you would probably do much better. We can't give you all day, but we can do the next best thing: We can give you more time for each question. How? By having you slow down and answer fewer questions.

Slowing down, and doing well on the questions you do answer, is the best way to improve your score on the AP Biology Exam. Rushing through questions in order to finish, on the other hand, will always hurt your score. When you rush, you're far more likely to make careless errors, misread, and fall into traps. Keep in mind that blank answers are not counted against you.

## THE THREE-PASS SYSTEM

The AP Biology Exam covers a broad range of topics. There's no way, even with our extensive review, that you will know everything about every topic in biology. So what should you do?

**3 is the Magic Number**
Using the Three-Pass System will help you earn more points by doing the questions you know first and saving the ones you don't know for last.

## Do the Easiest Questions First

The best way to rack up points is to focus on the questions you find the easiest. If you know the answer, nail it and move on. Other questions, however, will be a little more complicated. As you read each question, decide if you think it's easy, medium, or hard. During a first pass, do all the easy questions. If you come across a problem that seems time-consuming or completely incomprehensible, skip it. Remember:

> Questions that you find easy are worth just as many points as the ones you find difficult, so your time is better spent focusing on the ones in the former group.

Save the medium questions for the second pass. These questions are either time-consuming or require that you analyze all the answer choices (i.e., the correct answer doesn't pop off the page). If you come across a question that makes no sense from the outset, save it for the last pass. You're far more likely to fall into a trap or settle on a silly answer.

## Watch Out for Those Bubbles!

Since you're skipping problems, you need to keep careful track of the bubbles on your answer sheet. One way to accomplish this is by answering all the questions on a page and then transferring your choices to the answer sheet. If you prefer to enter them one by one, make sure you double-check the number beside the ovals before filling them in. We'd hate to see you lose points because you forgot to skip a bubble!

So then, what about the questions you don't skip?

# PROCESS OF ELIMINATION (POE)

On most tests, you need to know your material backward and forward to get the right answer. In other words, if you don't know the answer beforehand, you probably won't answer the question correctly. This is particularly true of fill-in-the-blank and essay questions. We're taught to think that the only way to get a question right is by knowing the answer. However, that's not the case on Section I of the AP Biology Exam. You can get a perfect score on this portion of the test without knowing a single right answer, provided you know all the wrong answers!

What are we talking about? This is perhaps the single most important technique in terms of the multiple-choice section of the exam. Let's take a look at the example below.

**The Art of POE**

You can earn a great score on the multiple-choice section by applying POE. Eliminating wrong answers can be just as effective as knowing the correct answer right off the bat.

> The structures that act as the sites of gas exchange in a woody stem are the
>
> (A)   lungs
>
> (B)   lenticels
>
> (C)   ganglia
>
> (D)   lentil beans

Now if this were a fill-in-the-blank-style question, you might be in a heap of trouble. But let's take a look at what we've got. You see "woody stem" in the question, which leads you to conclude that we're talking about plants. Right away, you know the answer is not (A) or (C) because plants don't have lungs or ganglia. Now we've narrowed it down to (B) and (D). Notice that (B) and (D) look very similar. Obviously, one of them is a trap. At this point, if you don't know what "lentil beans" are, you have to guess. However, even if we don't know precisely what they are, it's safe to say that most of us know that lentil beans have nothing to do with plant respiration. Therefore, the correct answer is (B), lenticels.

Although our example is a little goofy and doesn't look exactly like the questions you'll be seeing on the test, it illustrates an important point:

> **Process of Elimination (POE)** is the best way to approach the multiple-choice questions.

Even if you don't know the answer right off the bat, you'll surely know that two or three of the answer choices are not correct. What then?

## Aggressive Guessing

ETS, the service that develops and administers the exam, tells you that random guessing will not affect your score. This is true. There is no guessing penalty on the AP Biology Exam. For each correct answer you'll receive one point and you will not lose any points for each incorrect answer.

Although you won't lose any points for wrong answers, you should guess aggressively by getting rid of the incorrect answer choices. The moment you've eliminated a couple of answer choices, your odds of getting the question right, even if you guess, are far greater. If you can eliminate as many as two answer choices, your odds improve enough that it's in your best interest to guess.

## WORD ASSOCIATIONS

Another way to rack up the points on the AP Biology Exam is by using word associations in tandem with your POE skills. Make sure that you memorize the words in the Key Terms lists throughout this book. Know them backward and forward. As you learn them, make sure you group them by association, since ETS is bound to ask about them on the AP Biology Exam. What do we mean by "word associations"? Let's take the example of mitosis and meiosis.

You'll soon see from our review that there are several terms associated with mitosis and meiosis. *Synapsis, crossing-over*, and *tetrads*, for example, are words associated with meiosis but not mitosis. We'll explain what these words mean later in this book. For now, just take a look:

> Which of the following typifies cytokinesis
> during mitosis?
>
> (A)    Crossing-over
>
> (B)    Formation of tetrads
>
> (C)    Synapsis
>
> (D)    Division of the cytoplasm

This might seem like a difficult problem. But let's think about the associations we just discussed. The question asks us about mitosis. However, (A), (B), and (C) all mention events that we've associated with meiosis. Therefore, they are out. Without even racking your brain, you've managed to find the correct answer: (D). Not bad!

Once again, don't worry about the science for now. We'll review it later. What is important to recognize is that by combining the associations we'll offer throughout this book and your aggressive POE techniques, you'll be able to rack up points on problems that might have seemed difficult at first.

## MNEMONICS—OR THE BIOLOGY NAME GAME

One of the big keys to simplifying biology is the organization of terms into a handful of easily remembered packages. The best way to accomplish this is by using mnemonics. Biology is all about names: the names of chemical structures, processes, theories, and so on. How are you going to keep them all straight? A mnemonic, as you may already know, is a convenient device for remembering something.

For example, one important issue in biology is taxonomy, that is, the classification of life forms, or organisms. Organisms are classified in a descending system of similarity, leading from domains (the broadest level) to species (the most specific level). The complete order runs: domain, kingdom, phylum, class, order, family, genus, and species. Don't freak out yet. Look how easy it becomes with a mnemonic:

King Philip of Germany decided to walk to America. What do you think happened?

| | | |
|---|---|---|
| Dumb | → | Domain |
| King | → | Kingdom |
| Philip | → | Phylum |
| Came | → | Class |
| Over | → | Order |
| From | → | Family |
| Germany | → | Genus |
| Soaked | → | Species |

Learn the mnemonic and you'll never forget the science!

Mnemonics can be as goofy as you like, so long as they help you remember. Throughout this book, we'll give you mnemonics for many of the complicated terms we'll be seeing. Use ours, if you like them, or feel free to invent your own. Be creative! Remember: The important thing is that you remember the information, not how you remember it.

## IDENTIFYING EXCEPT QUESTIONS

About 10 percent of the multiple-choice questions in Section I are EXCEPT/ NOT/LEAST questions. With this type of question, you must remember that you're looking for the *wrong* (or the least correct) answer. The best way to go about these is by using POE.

More often than not, the correct answer is a true statement, but is wrong in the context of the question. However, the other three tend to be pretty straightforward. Cross off the three that apply and you're left with the one that does not. Here's a sample question:

> All of the following are true statements about gametes EXCEPT
>
> (A) They are haploid cells.
>
> (B) They are produced only in the reproductive structures.
>
> (C) They bring about genetic variation among offspring.
>
> (D) They develop from polar bodies.

If you don't remember anything about gametes and gametogenesis, or the production of gametes, this might be a particularly difficult problem. We'll see these again later on, but for now, remember that gametes are the "sex cells" of sexually reproducing organisms. As such, we know that they are haploid and are produced in the sexual organs. We also know that they come together to create offspring.

From this very basic review, we know immediately that (A) and (B) are not our answers. Both of these are accurate statements, so we eliminate them. That leaves us with (C) and (D). If you have no idea what (D) means, focus on (C). In sexual reproduction, each parent contributes one gamete, or half the genetic complement of the offspring. This definitely helps vary the genetic makeup of the offspring. Choice (C) is a true statement, so it can be eliminated. The correct answer is (D).

Don't sweat it if you don't recall the biology. We'll be reviewing it in detail soon enough. For now, remember that the best way to answer these types of questions is: Spot all the right statements and cross them off. You'll wind up with the wrong statement, which happens to be the correct answer

**Outsmart the Tricky Questions**
EXCEPT and NOT questions are considered more difficult than other question types. Here's where POE comes in handy!

## SOLVING GRID-IN QUESTIONS

Part B of the multiple-choice section requires you to answer six grid-in questions. Unlike the questions in Part A, you will not be able to use Process of Elimination. However, there are strategies you can use to increase your chances of getting these questions correct efficiently. First, identify what the question is actually asking for. Are you being asked to find a value on a graph or chart? Perhaps they are having you calculate the frequency of an allele using Hardy-Weinberg equilibrium. If you have no idea how to find out what they are asking you, take a quick educated guess and move on. For instance, if they are asking for a frequency, try guessing a decimal between 0 and 1, such as 0.25. However, if you do know how to find the answer, then write down the question number and any work (notes, calculations, etc.),  which you would need to answer the question neatly. Before filling in your answer, make sure that you are correct and haven't made any mistakes on interpreting data or in making calculations. Be sure not to forget to bubble in your answers.

**Remember Your Resources!**
The AP Biology formula sheets are included at the end of this book. Remember to use them when you take the Practice Tests.

Although these grid-in questions require calculations, fear not! The College Board provides you with a handy formula sheet, which you can find at the end of this book on page 381. You can expect to be given this sheet as part of the exam, so you do not need to memorize these equations. However, it will be important for you to practice using them and familiarizing yourself with the formula sheet prior to the exam. Keep this page handy as you review and practice problems, so that you will be accustomed to using it on the day of the exam.

# Chapter 2
# How to Approach
# Free-Response
# Questions

# THE ART OF THE FREE-RESPONSE ESSAY

On Section II of the exam, you are given two long-form free-response questions and six short-form free-response questions to answer in 80 minutes. You will also have a 10-minute reading period, which means you have a total of 90 minutes to complete Section II. The best way to rack up points on this section is to give the readers what they're looking for. Fortunately, we know precisely what that is.

The readers, or graders, have a checklist of key terms and concepts that they use to assign points. We like to call these "hot button" terms. Simply put, for each hot button term that you include in a response, you will receive a predetermined number of points. For example, if the question deals with the function of enzymes, the graders are instructed to give 2 points for a mention of the "lock-and-key theory of enzyme specificity."

Naturally, you can't just compose a "laundry list" of scientific terms. Otherwise, it wouldn't be an essay. What you can do, however, is organize your free-response essays around a handful of these key, or hot button, points. The most effective and efficient way to do this is by using the 10-minute reading period to brainstorm and come up with the scientific terms. Then outline your response before you begin to write, using your hot buttons as your guide.

# READ THE QUESTIONS CAREFULLY

**Make the Most of Your Planning Time**
If you use these 10 minutes wisely, you can breeze your way through the free-response questions.

ETS gives you 10 minutes to read the questions and organize your thoughts before you begin writing. The first thing you should do is take less than a minute to skim all of the questions and put them into your own personal order of difficulty from easiest to toughest. Once you've decided the order in which you will answer the questions (easiest first, hardest last), you can begin to formulate your responses. Your first step should be a more detailed assessment of each question.

The most important advice we can give you is to read each question at least twice. As you read the question, focus on key words, especially "direction words." Almost every essay question begins with a direction word. Some examples of direction words are *discuss*, *define*, *explain*, *describe*, *compare*, and *contrast*. If a question asks you to discuss a particular topic, you should give a viewpoint and support it with examples. If a question asks you to compare two things, you should discuss how the two things are similar. On the other hand, if the question asks you to contrast two things, you need to show how these things are different.

Many students lose points on their essays because they either misread the question or fail to do what's asked of them.

# BRAINSTORM

Your next objective during the 10-minute reading period should be to organize your thoughts.

Once you've read the questions, you need to brainstorm. Jot down as many key terms and concepts as you can. Don't forget, the test readers assign points on the basis of these key concepts. For each one that you mention and/or explain, you get a point.

How many do you need? You won't need all of them, that's for sure. However, you will need enough to get you the maximum number of points for that question. Most of these can be pulled directly from your reading in this book. At the end of each chapter is a Key Terms list which is full of terms that you should incorporate into your free responses. In addition to this list, you can feel free to create your own thorough lists of terms for your own use.

Don't spend more than about 2 minutes per question brainstorming. Once your 10-minute reading period is up, you should be ready to start writing your essays.

# OUTLINE YOUR RESPONSE

Have you ever written yourself into a corner? You're halfway through your essay when you suddenly realize that you have no idea whatsoever where you're going with your train of thought. To avoid this (and the panic that accompanies it), take a few minutes to draft an outline.

Your outline should incorporate as many of the hot buttons as you need in order to maximize your score. In other words, a question asks for two examples, choose only those two with which you are most comfortable. In your outline, make notes about the crucial points to mention with regard to each topic or key word. Once your outline is complete, you're ready to move on to writing the actual essay.

All of this preparation may seem time-consuming. However, it should take no more than four or five minutes per question. What's more, it will greatly simplify the whole essay-writing process. So while you lose a little time at the outset, you'll more than make up for it when it comes time to actually write your answer.

# DEVELOP YOUR IDEAS IN EACH RESPONSE

Now you can use your outline to write your essay. Most students can come up with key terms or phrases that concern a particular biology topic. What separates a high-scoring student from a low-scoring student is *how* the student develops his or her thoughts in each essay. Besides giving the hot buttons, you'll need to elaborate on your thoughts and ideas. For example, don't just throw out a list of terms that pertain to meiosis and mitosis (such as *synapsis*, *crossing-over*, and *gametes*). Go one step further. Make sure you mention the *significance* of meiosis (it produces genetic variability). This extra piece of information will earn you an extra point.

Generally, you'll need to write about two to four paragraphs, depending on the number of parts contained in each essay question. In addition, be sure to give the appropriate number of examples for each essay. If the question asks for three examples, give only three examples. If you present more than is required, the test readers won't even read them or count them toward your score. The bottom line is this: Stick to the question.

## ANSWER EACH PART OF THE QUESTION SEPARATELY

**Head over to the College Board**

For sample free-response questions and more, check out its AP Central page on the College Board website.

The more parts there are to a free-response question, the more important it is to pace yourself. For each free response, you're better off writing a little bit for each part than you are spending all your time on any one part of a question. Why? Even if you were to write the perfect answer to one part of a question, there's a limit to the number of points the test readers can assign to that part. Moreover, by writing a separate paragraph for each section, you make the test reader's job that much easier. When test readers have an easy time reading your essay, they're more likely to award points: It comes across as clearer and more organized. Readers have also requested that students label each part of their essay answers, so write "Part (a)," "Part (b)," and so on, accordingly in your response.

Finally, don't spend too much time writing a fancy introduction. You won't get brownie points for beautifully written openings like, "It was the best of experiments, it was the worst of experiments." Just leap right into the essay. And don't worry too much about grammar or spelling errors. Your grammar can hurt only if it's so bad that it seriously impairs your ability to communicate.

## INCORPORATE ELEMENTS OF AN EXPERIMENTAL DESIGN

Since one of the free-response questions will be experimentally based, you'll need to know how to design an experiment. Most of these questions require that you present an appropriately labeled diagram or graph. Otherwise, you'll only get partial credit for your work.

There are two things you must remember when designing experiments on the AP Biology Exam: (1) Always label your figures, and (2) include controls in all experiments. Let's take a closer look at these two points.

# Know How to Label Diagrams and Figures

Let's briefly discuss the important elements in setting up a graph. The favorite type of graph on the AP Biology Exam is the *coordinate graph*. The coordinate graph has a horizontal axis (*x*-axis) and a vertical axis (*y*-axis).

The *x*-axis usually contains the *independent variable*—the thing that's being manipulated or changed. The *y*-axis contains the *dependent variable*—the thing that's affected when the independent variable is changed.

Now let's look at what happens when you put some points on the graph. Every point on the graph represents both an independent variable and a dependent variable.

Once you draw both axes and label the axes as *x* and *y*, you can plot the points on the graph. Let's look at the following question.

Enzymes are biological catalysts.

(a) **Relate** the chemical structure of an enzyme to its catalytic activity and specificity.

(b) **Design** an experiment that investigates the influence of temperature, substrate concentration, or pH on the activity of an enzyme.

(c) **Describe** what information concerning enzyme structure could be inferred from the experiment you have designed.

For now, let's discuss only (b), which asks us to design an experiment. Let's set up a graph that shows the results of an experiment examining the relationship between pH and enzyme activity. Notice that we've chosen only one factor here, pH. We could have chosen any of the three. Why did we choose pH and not temperature or substrate concentration? Well, perhaps it's the one we know the most about.

What is the independent variable? It is pH. In other words, pH is being manipulated in the experiment. We'll therefore label the x-axis with pH values from 0 through 14.

What is the dependent variable? It's the enzyme activity—the thing that's affected by pH. Let's label the y-axis "Rate of Reaction." Now we're ready to plot the values on the graph. Based on our knowledge of enzymes, we know that for most enzymes the functional range of pH is narrow, with optimal performance occurring at or around a pH of 7.

Now you should interpret your graph. If the pH level decreases from a neutral pH of 7, the reaction rate of the enzyme will decrease. If the pH level increases, the rate of reaction will also decrease. Don't forget to include a simple explanation of your graph.

## INCLUDE CONTROLS IN YOUR EXPERIMENTS

Almost every experiment will have at least one variable that remains constant throughout the study. This is called the *control*. A control is simply a standard of comparison. What does a control do? It enables the biologist to verify that the outcome of the study is due to changes in the independent variable and nothing else.

Let's say the principal of your school thinks that students who eat breakfast do better on the AP Biology Exam than those who don't eat breakfast. He gives a group of 10 students from your class free breakfast every day for a year. When the school year is over, he administers the AP Biology Exam and they all score brilliantly! Did they do well because they ate breakfast every day? We don't know for sure. Maybe the principal handpicked the smartest kids in the class to participate in the study.

In this case, the best way to be sure that eating breakfast made a difference is to have a control group. In other words, he would need to pick students in the class who *never* eat breakfast and follow them for a year. At the end of that year, he could send them in to take the AP Biology Exam. If they do just as well as the group that ate breakfast, then we can probably conclude that eating breakfast wasn't the only factor leading to higher AP scores. The group of students that didn't eat breakfast is called the control group because those students were not "exposed" to the variable of interest—in this case, breakfast.

Let's see if you can write a good essay using your free-response techniques. Take about 22 minutes to write a response to the following free-response question:

> All organisms need nutrients to survive. Angiosperms and vertebrates have each developed various methods to obtain nutrients from their environment.
>
> (a) **Discuss** the ways angiosperms and vertebrates procure their nutrients.
>
> (b) **Discuss** two structures used for obtaining nutrients among angiosperms. **Relate** structure to function.
>
> (c) **Discuss** two examples of symbiotic relationships that have evolved between organisms to obtain nutrients.

**Put It All Together**
Use the techniques discussed in this chapter and practice writing your free responses. Additional sample free-response questions are available on the AP Central page for the AP Biology Exam.

# Chapter 3
# Using Time Effectively to Maximize Points

## BECOMING A BETTER TEST TAKER

Very few students stop to think about how to improve their test-taking skills. Most assume that if they study hard, they will test well, and if they do not study, they will do poorly. Most students continue to believe this even after experience teaches them otherwise. Have you ever studied really hard for an exam and then blown it on test day? Have you ever aced an exam for which you thought you weren't well prepared? Most students have had one, if not both, of these experiences. The lesson should be clear: Factors other than your level of preparation influence your final test score. This chapter will provide you with some insights that will help you perform better on the AP Biology Exam and on other exams as well.

## PACING AND TIMING

A big part of scoring well on an exam is working at a consistent pace. The worst mistake made by inexperienced or unsavvy test takers is that they come to a question that stumps them and rather than just skip it, they panic and stall. Time stands still when you're working on a question you cannot answer, and it is not unusual for students to waste five minutes on a single question (especially a question involving a graph or the word EXCEPT) because they are too stubborn to cut their losses. It is important to be aware of how much time you have spent on a given question and on the section you are working. There are several ways to improve your pacing and timing for the test.

- **Know your average pace.** While you prepare for your test, try to gauge how long you take on 5, 10, or 20 questions. Knowing how long you spend on average per question will help you identify how many questions you can answer effectively and how best to pace yourself for the test.
- **Have a watch or clock nearby.** You are permitted to have a watch or clock nearby to help you keep track of time. However, it's important to remember that constantly checking the clock is in itself a waste of time and can be distracting. Devise a plan. Try checking the clock every 15 or 30 questions to see if you are keeping the correct pace or whether you need to speed up. This will ensure that you're cognizant of the time but will not permit you to fall into the trap of dwelling on it.
- **Know when to move on.** Since all questions are scored equally, investing appreciable amounts of time on a single question is inefficient and can potentially deprive you of the chance to answer easier ones later on. You should eliminate answer choices if you are able to, but don't worry about picking a random answer and moving on if you cannot find the correct answer. Remember, tests are like marathons; you do best when you work through them at a steady pace. You can always come back to a question you don't know. When you do, very

often you will find that your previous mental block is gone, and you will wonder why the question perplexed you the first time around (as you gleefully move on to the next question). Even if you still don't know the answer, you will not have wasted valuable time you could have spent on easier questions.

- **Be selective.** You don't have to do any of the questions in a given section in order. If you are stumped by an essay or multiple choice question, skip it or choose a different one. In the next section, you will see that you may not have to answer every question correctly to achieve your desired score. Select the questions or essays that you can answer and work on them first. This will make you more efficient and give you the greatest chance of getting the most questions correct.
- **Use Process of Elimination (POE) on multiple-choice questions.** Many times, one or more answer choices can be eliminated. Every answer choice that can be eliminated increases the odds that you will answer the question correctly.

Remember, when all the questions on a test are of equal value, no one question is that important, and your overall goal for pacing is to get the most questions correct. Finally, you should set a realistic goal for your final score. In the next section, we will break down how to achieve your desired score and how to pace yourself to do so.

## GETTING THE SCORE YOU WANT

Depending on the score you need, it may be in your best interest not to try to work through every question. Check with the schools to which you are applying.

AP exams in all subjects no longer include a "guessing penalty" of a quarter of a point for every incorrect answer. Instead, students are assessed only on the total number of correct answers. A lot of AP materials, even those you receive in your AP class, may not include this information. It's really important to remember that if you are running out of time, you should fill in all the bubbles before the time for the multiple-choice section is up. Even if you don't plan to spend a lot of time on every question or even if you have no idea what the correct answer is, you need to fill something in.

# TEST ANXIETY

Everybody experiences anxiety before and during an exam. To a certain extent, test anxiety can be helpful. Some people find that they perform more quickly and efficiently under stress. If you've ever pulled an all-nighter to write a paper and ended up doing good work, you know the feeling.

However, too much stress is definitely a bad thing. Hyperventilating during the test, for example, almost always leads to a lower score. If you find that you stress out during exams, here are a few preemptive actions you can take.

- **Take a reality check.** Evaluate your situation before the test begins. If you have studied hard, remind yourself that you are well prepared. Remember that many others taking the test are not as well prepared, and (in your classes, at least) you are being graded against them, so you have an advantage. If you didn't study, accept the fact that you will probably not ace the test. Make sure you get to every question you know something about. Don't stress out or fixate on how much you don't know. Your job is to score as high as you can by maximizing the benefits of what you do know. In either scenario, it's best to think of a test as if it were a game. How can you get the most points in the time allotted to you? Always answer questions you can answer easily and quickly before tackling those that will take more time.
- **Try to relax.** Slow, deep breathing works for almost everyone. Close your eyes, take a few, slow, deep breaths, and concentrate on nothing but your inhalation and exhalation for a few seconds. This is a basic form of meditation that should help you to clear your mind of stress and, as a result, concentrate better on the test. If you have ever taken yoga classes, you probably know some other good relaxation techniques. Use them when you can (obviously, anything that requires leaving your seat and, say, assuming a handstand position won't be allowed by any but the most free-spirited proctors).
- **Eliminate as many surprises as you can.** Make sure you know where the test will be given, when it starts, what type of questions are going to be asked, and how long the test will take. You don't want to be worrying about any of these things on test day or, even worse, after the test has already begun.

The best way to avoid stress is to study both the test material and the test itself. Congratulations! By buying or reading this book, you are taking a major step toward a stress-free AP Biology Exam.

# REFLECT

Respond to the following questions:

- How long will you spend on multiple-choice questions?

- How will you change your approach to multiple-choice questions?

- What is your multiple-choice guessing strategy?

- How much time will you spend on the free-response questions?

- What will you do before you begin writing your free-response answers?

- Will you seek further help, outside of this book (such as a teacher, tutor, or AP Central), on how to approach the questions that you will see on the AP Biology Exam?

# Part V
# Content Review for the AP Biology Exam

# Chapter 4
# The Chemistry
# of Life

## ELEMENTS

Although organisms exist in many diverse forms, they all have one thing in common: They are all made up of matter. Matter is made up of elements. **Elements**, by definition, are substances that cannot be broken down into simpler substances by chemical means.

## THE ESSENTIAL ELEMENTS OF LIFE

Although there are 92 natural elements, 96 percent of the mass of all organisms is made up of just four of them: **oxygen** (O), **carbon** (C), **hydrogen** (H), and nitrogen (N). Other elements such as calcium (Ca), phosphorus (P), potassium (K), and magnesium (Mg) are also present, but in smaller quantities. These elements make up most of the remaining 4 percent of an organism's weight. Some elements are known as **trace elements** because they are only required by an organism in very small quantities. Trace elements include iron (Fe), iodine (I), and copper (Cu).

## SUBATOMIC PARTICLES

If you break down an element into smaller pieces, you'll eventually come to the **atom**—the smallest unit of an element that retains its characteristic properties. Atoms are the building blocks of the physical world.

Within atoms, there are even smaller subatomic particles called **protons**, **neutrons**, and **electrons**. Let's take a look at a typical atom:

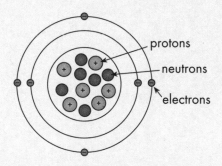

Protons and neutrons are particles that are packed together in the core of an atom called the **nucleus**. You'll notice that protons are positively charged (+) particles, whereas neutrons are uncharged particles.

Electrons, on the other hand, are negatively charged (–) particles that spin around the nucleus. Electrons are pretty small compared to protons and neutrons. In fact, for our purposes, electrons are considered massless. Most atoms have the same number of protons and electrons, making them electrically neutral. Some atoms have the same number of protons but differ in the number of neutrons in the nucleus. These atoms are called **isotopes**.

**Carbon 14 Isotopes**
Ancient artifacts can be dated by examining the rate of decay in carbon 14 isotopes.

## COMPOUNDS

When two or more different types of atoms are combined in a fixed ratio, they form a chemical **compound**. You'll sometimes find that a compound has different properties from those of its elements. For instance, hydrogen and oxygen exist in nature as gases. Yet when they combine to make water, they often pass into a liquid state. When hydrogen atoms get together with oxygen atoms to form water, we've got a **chemical reaction**:

$$2H_2 \, (g) + O_2 \, (g) \rightarrow 2H_2O \, (l)$$

The atoms of a compound are held together by **chemical bonds**, which may be **ionic bonds**, **covalent bonds**, or **hydrogen bonds**.

An ionic bond is formed between two atoms when one or more electrons are transferred from one atom to the other. In this reaction, one atom *loses* electrons and becomes positively charged, and the other atom *gains* electrons and becomes negatively charged. The charged forms of the atoms are called **ions**. An ionic bond results from the attraction between two oppositely charged ions. For example, when Na reacts with Cl, charged ions, $Na^+$ and $Cl^-$, are formed.

A covalent bond is formed when electrons are *shared* between atoms. If the electrons are shared equally between the atoms, the bond is called **nonpolar covalent**. If the electrons are shared unequally, the bond is called **polar covalent**. When one pair of electrons is shared between two atoms, the result is a single covalent bond. When two pairs of electrons are shared, the result is a double covalent bond. When three pairs of electrons are shared, the result is a triple covalent bond.

## WATER: THE VERSATILE MOLECULE

One of the most important substances in nature is water. Water is considered a unique molecule because it plays an important role in chemical reactions.

Let's take a look at one of the properties of water. Water has two hydrogen atoms joined to an oxygen atom:

**Water Weight**
Did you know that more than 60 percent of your body weight consists of water?

In water molecules, the hydrogen atoms have a partial positive charge and the oxygen atom has a partial negative charge. Molecules that have partially positive and partially negative charges are said to be **polar**. Water is therefore a polar molecule. The positively-charged ends of the water molecules strongly attract the negatively-charged ends of other polar compounds. Likewise, the negatively-charged ends strongly attract the positively-charged ends of neighboring compounds. These forces are most readily apparent in the tendency of water molecules to stick together, as in the formation of water beads or raindrops.

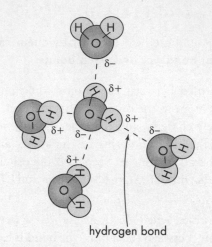

hydrogen bond

These types of intermolecular attractions are called hydrogen bonds. Hydrogen bonds are weak chemical bonds that form when a hydrogen atom that is covalently bonded to one electronegative atom is also attracted to another electronegative atom. Water molecules are held together by hydrogen bonds. Although hydrogen bonds are individually weak, they are strong when present in large numbers. Because it can react with other polar substances, water makes a great solvent; it can dissolve many kinds of substances. The hydrogen bonds that hold water molecules together contribute to a number of special properties, including **cohesion**, **adhesion**, **heat capacity**, and **expansion on freezing**.

## Cohesion

As mentioned above, water molecules have a strong tendency to stick together. That is, water exhibits *cohesive forces*. These forces are extremely important to life. For instance, during transpiration water molecules evaporate from a leaf, "pulling" on neighboring water molecules. These, in turn, draw up the molecules immediately behind them, and so on, all the way down the plant vessels. The resulting chain of water molecules enables water to move up the stem. Water also has a high surface tension because of the cohesiveness of its molecules.

## Adhesion

Water molecules also like to stick to other substances—that is, they're *adhesive*. Have you ever tried to separate two glass slides stuck together by a film of water? They're difficult to separate because of the water sticking to the glass surfaces. These two forces taken together—**cohesion** and **adhesion**—account for the ability of water to rise up the roots, trunks, and branches of trees. Since this phenomenon occurs in thin vessels, it's called **capillary action**.

## Heat Capacity

Another remarkable property of water is its high **heat capacity**. What's heat capacity? Your textbook will give you a definition something like this: "Heat capacity is the quantity of heat required to change the temperature of a substance by 1 degree." What does that mean? In plain English, heat capacity refers to the ability of a substance to store heat. For example, when you heat up an iron kettle, it gets hot pretty quickly. Why? Because it has a *low* specific heat. It doesn't take much heat to increase the temperature of the kettle. Water, on the other hand, has a high heat capacity. You have to add a lot of heat to get an increase in temperature. Water's ability to resist temperature changes is one of the things that helps keep the temperature in our oceans fairly stable. It's also why organisms that are mainly made up of water, like us, are able to keep a constant body temperature.

## Expansion on Freezing

Perhaps one of the most important properties of water is yet another result of hydrogen bonding. When four water molecules are bound in a solid lattice of ice, the hydrogen bonds actually cause solid water to expand on freezing. In most materials, when the molecules lose kinetic energy and cool from liquid to solid, the molecules get more dense, moving closer together. However, in liquid water, the molecules are slightly more dense than in solid water. Because water expands on freezing, becoming slightly less dense than liquid water, ice can crack lead pipes in the winter, or a soda can will pop if it is left in your car. The important consequence of this property of water is that ice floats on the top of lakes or streams, allowing animal life to live underneath the ice. If ice was denser than water, it would sink to the bottom freezing the body of water solid. All aquatic life would cease.

So let's review the unique properties of water.

- Water is polar and can dissolve other polar substances.

- Water has cohesive and adhesive properties.

- Water has a high heat capacity.

- Water has a high **surface tension**.

- Water expands when it freezes (or, ice floats).

# THE ACID TEST

We just said that water is important because most reactions occur in watery solutions. Well, there's one more thing to remember: Reactions are also influenced by whether the solution in which they occur is **acidic**, **basic**, or **neutral**.

What makes a solution acidic or basic? A solution is acidic if it contains a lot of *hydrogen ions* ($H^+$). That is, if you dissolve an acid in water, it will release a lot of hydrogen ions. When you think about acids, you usually think of substances with a sour taste, like lemons. For example, if you squeeze a little lemon juice into a glass of water, the solution will become acidic. That's because lemons contain citric acid.

Bases, on the other hand, do not release hydrogen ions when added to water. They release a lot of *hydroxide ions* ($OH^-$). These solutions are said to be **alkaline**. Bases usually have a slippery consistency. Common soap, for example, is composed largely of bases.

The acidity or alkalinity of a solution can be measured using a **pH scale**. The pH scale is numbered from 1 to 14. The midpoint, 7, is considered neutral pH. The concentration of hydrogen ions in a solution will indicate whether it is acidic, basic, or neutral. If a solution contains a lot of hydrogen ions, then it will be acidic and have a low pH. Here's the trend:

> An increase in $H^+$ ions causes a decrease in the pH.
>
> $$pH = -\log [H+]$$

Therefore, as the concentration of $H^+$ ions increases by a factor of 10, the pH becomes a smaller number. For example, stomach acid has a pH of 2, and if we use the equation, we will see that that means that the concentration of $H^+$ ions is 10–2 M, which is pretty high when you consider the other extreme is lye, which has a pH of 14, a concentration of $H^+$ ions around 10–14 M! Use your calculator to convince yourself that these numbers are correct.

You'll notice from the scale on the next page that stronger acids have lower pHs. If a solution has a low concentration of hydrogen ions, it will have a high pH.

| | | |
|---|---|---|
| More acidic | 0 | Concentrated nitric acid |
| pH | 1 | Stomach acid |
| | 2 | Lemon juice |
| | 3 | Cola drinks |
| | 4 | Vinegar |
| | 5 | Beer |
| | 6 | Black coffee |
| Neutral | 7 | Distilled water |
| | 8 | Blood |
| | 9 | Seawater |
| | 10 | Laundry bleach |
| | 11 | Ammonia |
| | 12 | |
| pH | 13 | Oven cleaner |
| More basic | 14 | Drain cleaner |

One more thing to remember: The pH scale is not a linear scale—it's logarithmic. That is, a change of *one* pH number actually represents a *tenfold* change in hydrogen ion concentration. For example, a pH of 3 is actually ten times more acidic than a pH of 4. This is also true in the reverse direction: A pH of 4 represents a tenfold decrease in acidity compared to a pH of 3.

# ORGANIC MOLECULES

Now that we've discussed chemical compounds in general, let's talk about a special group of compounds. Most of the chemical compounds in living organisms contain a skeleton of carbon atoms. These molecules are known as **organic compounds**. By contrast, molecules that do not contain carbon atoms are called **inorganic compounds**. For example, salt (NaCl) is an inorganic compound.

Carbon is important for life because it is a versatile atom, meaning that it not only has the ability to bind with other carbons but also with a number of other atoms including nitrogen, oxygen, and hydrogen. The resulting molecules are key in carrying out the activities necessary for life.

To recap:

- Organic compounds contain carbon atoms.

- Inorganic compounds do not contain carbon atoms (except carbon dioxide).

Now let's focus on four classes of organic compounds central to life on Earth:

- carbohydrates
- proteins
- lipids
- nucleic acids

## Carbohydrates

Organic compounds that contain carbon, hydrogen, and oxygen are called **carbohydrates**. They usually contain these three elements in a ratio of 1:2:1, respectively. We can represent the proportion of elements within carbohydrate molecules by the formula $C_nH_{2n}O_n$, which can be simplified as $(CH_2O)_n$.

Most carbohydrates are categorized as either **monosaccharides**, **disaccharides**, or **polysaccharides**. The term *saccharides* is a fancy word for "sugar." The prefixes *mono-*, *di-*, and *poly-* refer to the number of sugars in the molecule. *Mono-* means "one," *di-* means "two," and *poly-* means "many." A monosaccharide is therefore a carbohydrate made up of a single type of sugar molecule.

## Monosaccharides: The Simplest Sugars

Monosaccharides, the simplest sugars, serve as an energy source for cells. The two most common sugars are (1) **glucose** and (2) **fructose**.

Both of these monosaccharides are six-carbon sugars with the chemical formula $C_6H_{12}O_6$. Glucose, the most abundant monosaccharide, is the most popular sugar around. Plants produce it by capturing sunlight for energy, while cells break it down to release stored energy. Glucose can come in two forms: α-glucose and β–glucose, which differ simply by a reversal of the H and OH on the first carbon. Fructose, the other monosaccharide you need to know for the test, is a common sugar in fruits.

Glucose and fructose can be depicted as either "straight" or "rings." Both of them are pretty easy to spot; just look for the carbon molecules with a lot of OH's and H's attached to them.

Here are the two different forms:

Ring form of glucose

Straight-chain
form of glucose

Ring form of fructose

Straight-chain
form of fructose

## Disaccharides

What happens when two monosaccharides are brought together? The hydrogen (–H) from one sugar molecule combines with the hydroxyl group (–OH) of another sugar molecule. What do H and OH add up to? Water ($H_2O$)! So a water molecule is removed from the two sugars. The two molecules of monosaccharides are chemically linked and form a disaccharide.

Maltose is an example of a disaccharide:

Glucose          Glucose

Maltose

Maltose is formed by linking two glucose molecules—forming a **glycosidic bond**. This process is called **dehydration synthesis**, or **condensation**. During this process, a water molecule is lost.

Now, what if you want to break up the disaccharide and form two monosaccharides again? Just add water. That's called **hydrolysis**, which means "water" (*hydro-*) and "breaking" (*-lysis*). The water breaks the bond between the two monomers. Hydrolysis is a common reaction in biology.

Maltose

Glucose          Glucose

## Polysaccharides

Polysaccharides are made up of many repeated units of monosaccharides. Therefore, a polysaccharide is a kind of **polymer**, a molecule with repeating subunits of the same general type. The most common polysaccharides you'll need to know for the test are **starch**, **cellulose**, and **glycogen**. Polysaccharides are often storage forms of sugar or structural components of cells. For instance, animals store glucose molecules in the form of glycogen in the liver and muscle cells. Plants "stockpile" α–glucose in the form of starch in structures called **plastids**. Cellulose, on the other hand, made up of β–glucose, is a major part of the cell wall in plants. Its function is to lend structural support. Chitin, a polymer of β–glucose molecules, serves as a structural molecule in the walls of fungus and in the exoskeletons of arthropods.

Here's an AP question that may come up on the test: Why can't humans digest cellulose? The glycosidic bond in polymers that have α–glucose can easily be broken down by humans but glycoside bonds in polymers containing β–glucose polymers, such as cellulose, cannot. This is because the bonds joining the glucose subunits in cellulose are different than those in starch. Starch is composed of α–glucose subunits held together by 1,4 glycoside linkages while cellulose contain β–glucose subunits held together by 1,4 linkages.

## Proteins

**Amino acids** are organic molecules that serve as the building blocks of proteins. They contain carbon, hydrogen, oxygen, and nitrogen atoms. There are 20 different amino acids commonly found in proteins. Fortunately, you don't have to memorize the 20 amino acids. But you do have to remember that every amino acid has four important parts: an **amino group** (–NH$_2$), a **carboxyl group** (–COOH), a hydrogen, and an **R group**.

Here's a typical amino acid:

Amino acids differ only in the R group, which is also called the **side chain**. The R group associated with an amino acid could be as simple as a hydrogen atom (as in the amino acid *glycine*) or as complex as a carbon skeleton (as in the amino acid *arginine*).

Glycine                                    Arginine

Since ETS will probably test you on chemical diagrams, it's a good idea to identify the functional groups of the different structures. **Functional groups** are the distinctive groups of atoms that play a large role in determining the chemical behavior of the compound they are a part of. For example, an organic acid has a functional group, the carboxyl group (–COOH) that releases hydrogen ions in water. This makes the solution acidic. When it comes to spotting an amino acid, simply keep an eye out for the amino group ($NH_2$), then look for the carboxyl molecule (COOH). The most common functional groups of organic compounds are listed on the next page.

| Functional Group | Structural Formula | Molecular Formula |
|---|---|---|
| Amino | $\begin{matrix} & H \\ -&N-H \end{matrix}$ | $-NH_2$ |
| Alkyl | $\begin{matrix} H & H & & H \\ -C & -C- & \cdots & -C-H \\ H & H & & H \end{matrix}$ | $-C_nH_{2n+1}$ |
| Methyl | $\begin{matrix} H \\ -C-H \\ H \end{matrix}$ | $-CH_3$ |
| Ethyl | $\begin{matrix} H & H \\ -C-&C-H \\ H & H \end{matrix}$ | $-C_2H_5$ |
| Propyl | $\begin{matrix} H & H & H \\ -C-&C-&C-H \\ H & H & H \end{matrix}$ | $-C_3H_7$ |
| Carboxyl | $\begin{matrix} O \\ \parallel \\ -C-O-H \end{matrix}$ | $-COOH$ |
| Hydroxyl | $-O-H$ | $-OH$ |
| Aldehyde | $\begin{matrix} O \\ \parallel \\ -C-H \end{matrix}$ | $-CHO$ |
| Keto (carbonyl) | $\begin{matrix} O \\ \parallel \\ -C- \end{matrix}$ | $-CO$ |
| Sulfhydryl (thiol) | $-S-H$ | $-SH$ |
| Phenyl | (ring structure) | $-C_6H_5$ |
| Phosphate | $\begin{matrix} O \\ \parallel \\ -O-P-O \\ \mid \\ O \end{matrix}$ | $-PO_4$ |

## Polypeptides

When two amino acids join they form a **dipeptide**. The carboxyl group of one amino acid combines with the amino group of another amino acid. Here's an example:

Here's the peptide bond

This is the same process we saw earlier: dehydration synthesis. Why? Because a water molecule is removed to form a bond. By the way, the bond between two amino acids has a special name—a **peptide bond**. If a group of amino acids are joined together in a "string," the resulting organic compound is called a **polypeptide**. Once a polypeptide chain twists and folds on itself, it forms a three-dimensional structure called a **protein**.

# Lipids

Like carbohydrates, **lipids** consist of carbon, hydrogen, and oxygen atoms, but not in the 1:2:1 ratio typical of carbohydrates. The most common examples of lipids are **fats**, **oils**, **phospholipids**, and **steroids**. Let's talk about the simple lipids—**neutral fats**. A typical fat consists of three fatty acids and one molecule of **glycerol**. If you see the word *triglyceride* on the test, it's just a fancy word for "fat." Let's take a look:

Glycerol ——

Ester Linkage

To make a triglyceride, each of the carboxyl groups (–COOH) of the three fatty acids must react with one of the three hydroxyl groups (–OH) of the glycerol molecule. This happens by the removal of a water molecule. So, the creation of a fat requires the removal of three molecules of water. Once again, what have we got? You probably already guessed it—dehydration synthesis! The linkage now formed between the glycerol molecule and the fatty acids is called **ester linkage**. A fatty acid can be **saturated**, which means it has a single covalent bond between each pair of carbon atoms, or it can be **unsaturated**, which means adjacent carbons are joined by double bonds instead of single bonds. A **polyunsaturated** fatty acid has many double bonds within the fatty acid.

Lipids are important because they function as structural components of cell membranes, sources of insulation, and a means of energy storage.

## Phospholipids

Another special class of lipids is known as phospholipids. Phospholipids contain two fatty acid "tails" and one negatively charged phosphate "head." Take a look at a typical phospholipid:

glycerol                                    fatty acids

Phospholipids are extremely important, mainly because of some unique properties they possess, particularly with regard to water.

Interestingly enough, the two fatty acid tails are **hydrophobic** ("water-hating"). In other words, just like oil and vinegar, fatty acids and water don't mix. The reason for this is that fatty acid tails are nonpolar, and nonpolar substances don't mix well with polar ones, such as water.

On the other hand, the phosphate "head" of the lipid is **hydrophilic** ("water-loving"), meaning that it does mix well with water. Why? It carries a negative charge, and this charge draws it to the positively charged end of a water molecule. A molecule is **amphipathic** if it has both a hydrophilic region and a hydrophobic region. A phospholipid is amphipathic.

This arrangement of the fatty acid tails and the phosphate group head provides phospholipids with a unique shape. The two fatty acid chains orient themselves away from water, while the phosphate portion orients itself toward the water. Keep these properties in mind. We'll see later how this orientation of phospholipids in water relates to the structure and function of cell membranes.

One class of lipids is known as steroids. All steroids have a basic structure of four linked carbon rings. This category includes cholesterol, vitamin D, and a variety of hormones. Take a look at a few typical steroids:

**Estradiol**

**Testosterone**

**Cholesterol**

## Nucleic Acids

The fourth class of organic compounds are the **nucleic acids**. Like proteins, nucleic acids contain carbon, hydrogen, oxygen, and nitrogen, but nucleic acids also contain phosphorus. Nucleic acids are molecules that are made up of simple units called **nucleotides**. For the AP Biology Exam, you'll need to know about two kinds of nucleic acids: **deoxyribonucleic acid (DNA)** and **ribonucleic acid (RNA)**.

DNA is important because it contains genes, the hereditary "blueprints" of all life. RNA is important because it's essential for protein synthesis. We'll discuss DNA and RNA in greater detail when we discuss heredity. The following table summarizes the important macromolecules discussed here.

**Important Biological Macromolecules**

| Macromolecule | Monomer | Polymer | Linkage Bond |
|---|---|---|---|
| Carbohydrate | Monosaccharide (ex. Glucose) | Polysaccharide (ex. Starch, glycogen, cellulose) | Glycosidic Linkage |
| Protein | Amino Acid (ex. Glycine) | Polypeptide (ex. Actin) | Peptide Bond |
| Nucleic Acids | Nucleotides (ex. Adenine, thymine, guanine, cytosine) | Deoxyribonucleic acid (DNA) Ribonucleic acid (RNA) | Sugar-phosphate phosphodiester bonds |
| Lipids | (Not a true polymer.) | Compose fats, oils, phospholipids, steroids (ex. triglycerides) | Ester bonds |

# THE HETEROTROPH HYPOTHESIS

We've just seen the organic compounds that are essential for life. But where did they come from in the first place? This is still a hotly debated topic among scientists. Most scientists believe that the earliest precursors of life arose from nonliving matter (basically gases) in the primitive oceans of the earth. But this theory didn't take shape until the 1920s. Two scientists, **Alexander Oparin** and **J. B. S. Haldane**, proposed that the primitive atmosphere contained the following gases: methane ($CH_4$), ammonia ($NH_3$), hydrogen ($H_2$), and water ($H_2O$). Interestingly enough, there was almost no free oxygen ($O_2$) in this early atmosphere. They believed that these gases collided, producing chemical reactions that eventually led to the organic molecules we know today.

This theory didn't receive any substantial support until 1953. In that year, **Stanley Miller** and **Harold Urey** simulated the conditions of primitive Earth in a laboratory. They put the gases theorized to be abundant in the early atmosphere into a flask, struck them with electrical charges in order to mimic lightning, and organic compounds similar to amino acids appeared!

But how do we make the leap from simple organic molecules to more complex compounds and life as we know it? Since no one was around to witness the process, no one knows for sure how (or when) it occurred. Complex organic compounds (such as proteins) must have formed via dehydration synthesis. Simple cells then used organic molecules as their source of food. Over time, simple cells evolved into complex cells.

Now let's throw in a few new terms. Living organisms that rely on organic molecules for food are called **heterotrophs** (or consumers). For example, we're heterotrophs. Eventually, some organisms (such as plants) found a way to make their own food. These organisms are called **autotrophs** (or producers). Early autotrophs are responsible for Earth's oxygenated atmosphere.

# KEY TERMS

elements
oxygen
carbon
hydrogen
nitrogen
trace elements
atom
protons
neutrons
electrons
nucleus
isotopes
compound
chemical reaction
chemical bond
ionic bond
ions
covalent bond
nonpolar covalent
polar covalent
polar
hydrogen bonds
cohesion
adhesion
capillary action
heat capacity
surface tension
acidic
basic
neutral
alkaline
pH scale
organic compounds
inorganic compounds
carbohydrates
monosaccharides
disaccharides
polysaccharides
glucose
fructose
glycosidic bond

dehydration synthesis
(or condensation)
hydrolysis
polymer
starch
cellulose
glycogen
plastids
amino acids
amino group
carboxyl group
R group
side chain
functional groups
dipeptide
peptide bond
polypeptide
protein
lipids
fats
oils
phospholipids
steroids
neutral fats
glycerol
ester linkage
saturated
unsaturated
polyunsaturated
hydrophobic
hydrophilic
amphipathic
nucleic acids
nucleotides
deoxyribonucleic acid (DNA)
ribonucleic acid (RNA)
Oparin and Haldane
Stanley Miller
Harold Urey
heterotrophs (or consumers)
autotrophs (or producers)

# Summary

- Compounds are composed of two or more types of elements in a fixed ratio.

- Compounds are held together by ionic bonds and covalent bonds.

- Hydrogen bonding is essential for these emergent properties of water:
  - cohesion
  - adhesion
  - high heat capacity
  - expansion upon freezing
  - polar, and can dissolve other polar substances

- pH is important for biological reactions. Solutions may be acidic, basic or alkaline, or neutral.

- Organic molecules contain carbon atoms. Many molecules in biology are composed of several monomers bound together into polymers. The creation of these polymers occurs through dehydration synthesis reactions. The breakdown of these polymers occurs through hydrolysis reactions.

- Important organic macromolecules in biology include:
  - carbohydrates
  - proteins
  - nucleic acids
  - lipids

# Chapter 4 Drill

Answers and explanations can be found in Chapter 16.

1. Water is a critical component of life due to its unique structural and chemical properties. Which of the following does NOT describe a way that the exceptional characteristics of water are used in nature for life?

   (A) The high heat capacity of water prevents lakes and streams from rapidly changing temperature and freezing completely solid in the winter.

   (B) The high surface tension and cohesiveness of water facilitates capillary action in plants.

   (C) The low polarity of water prevents dissolution of cells and compounds.

   (D) The high intermolecular forces of water, such as hydrogen bonding, results in a boiling point which exceeds the tolerance of most life on the planet.

2. The intracellular pH of human cells is approximately 7.4. Yet, the pH within the lumen (inside) of the human stomach averages 1.5. Which of the following accurately describes the difference between the acidity of the cellular and gastric pH?

   (A) Gastric juices contain approximately 5 times more H+ ions than the intracellular cytoplasm of cells and are more acidic.

   (B) Gastric juices contain approximately 100,000-fold more H+ ions than the intracellular cytoplasm of cells and are more acidic.

   (C) The intracellular cytoplasm of cells contain approximately 5 times more H+ ions than gastric juices and is more acidic.

   (D) The intracellular cytoplasm of cells contains approximately 100,000-fold more H+ ions than gastric juices and is more acidic.

3. Amino acids are the basic molecular units which comprise proteins. All life on the planet forms proteins by forming chains of amino acids. Which labeled component of the amino acid structure of phenylalanine shown below will vary from amino acid to amino acid?

Questions 4–6 refer to the following paragraph and diagram.

In 1953, Stanley Miller and Harold Urey performed an experiment at the University of Chicago to test the hypothesis that the conditions of the early Earth would have favored the formation of larger, more complex organic molecules from basic precursors. The experiment, as shown below, consisted of sealing basic organic chemicals (representing the atmosphere of the primitive Earth) in a flask, which was exposed to electric sparks (to simulate lightning) and water vapor.

After one day of exposure, the mixture in the flask had turned pink in color, and later analysis showed that at least 10% of the carbon had been transformed into simple and complex organic compounds including at least 11 different amino acids and some basic sugars. No nucleic acids were detected in the mixture.

4. Which of the following contradicts the hypothesis of the experiment that life may have arisen from the formation of complex molecules in the conditions of the primitive Earth?

   (A) Complex carbon-based compounds were generated after only one day of exposure to simulated primitive Earth conditions.
   (B) Nucleic acid compounds such as DNA and RNA were not detected in the mixture during the experiment.
   (C) Over half of known amino acids involved in life were detected in the mixture during the experiment.
   (D) Basic sugar molecules were generated and detected in the mixture during the experiment.

5. Some amino acids, such as cysteine (shown below) and methionine, could not be formed in this experiment. Which of the following best explains why these molecules could not be detected?

   (A) The chemical reactions necessary to create amino acids such as cysteine and methionine require more energy than the simulated lightning provided in the experiment.
   (B) The chemical reactions necessary to create amino acids such as cysteine and methionine require enzymes for catalysis to occur, which were not included in the experiment.
   (C) Sulfur-based compounds were not included in the experiment.
   (D) Nitrogen-based compounds were not included in the experiment.

6. A scientist believes that the Miller-Urey experiment was unable to capture the remaining amino acids, and the formation of nucleic acids was due to the absence of critical chemical substrates that would have existed on the primordial Earth due to volcanism. All of the following basic compounds which are associated with volcanism should be included by the scientist EXCEPT

   (A) $H_2S$ gas
   (B) $SiO_2$ (silica)
   (C) $SO_2$
   (D) $H_3PO_4$ (phosphoric acid)

7. Catabolism refers to breaking down complex macromolecules into their basic components. Many biological processes use hydrolysis for catabolism. Hydrolysis of proteins could directly result in

(A) free water
(B) adenine
(C) cholesterol
(D) dipeptides

8. Which of the following contain both hydrophilic and hydrophobic properties and are often found in cell plasma membranes?

(A) Nucleotides
(B) Phospholipids
(C) Water
(D) Amino acids

9. Maltotriose is a trisaccharide composed of three glucose molecules linked through α-1,4 glyco-sidic linkages. Maltotriose frequently results from the hydrolysis of starch by the enzyme amylase in human saliva. What would the formula be for maltotriose?

(A) $C_{18}H_{36}O_{18}$
(B) $C_{18}H_{10}O_{15}$
(C) $C_{18}H_{32}O_{16}$
(D) $C_3H_6O_3$

10. A radioactive isotope of hydrogen, 3H, is called tritium. Tritium differs from the more common form of hydrogen because

(A) it contains two neutrons and one proton in its nucleus
(B) it contains one neutron and two protons in its nucleus
(C) it differs by its atomic number
(D) it is radioactive and therefore gives off one electron

# REFLECT

Respond to the following questions:

- For what content topics discussed in this chapter do you feel you have achieved sufficient mastery to answer multiple-choice questions correctly?

- For which content topics discussed in this chapter do you feel you have achieved sufficient mastery to discuss effectively in a free response?

- For which content topics discussed in this chapter do you feel you need more work before you can answer multiple-choice questions correctly?

- For which content topics discussed in this chapter do you feel you need more work before you can discuss effectively in a free response?

- What parts of this chapter are you going to re-review?

- Will you seek further help, outside this book (such as a teacher, tutor, or AP Central), on any of the content in this chapter—and if so, on what content?

# Chapter 5
# Cells

## LIVING THINGS

All living things are composed of **cells**. According to the cell theory, the cell is life's basic unit of structure and function. This simply means that the cell is the smallest unit of living material that can carry out all the activities necessary for life.

### Microscopes and the Study of Cells

Cells are studied using different types of microscopes. **Light microscopes**, also known as compound microscopes, are used to study stained or living cells. They can magnify the size of an organism up to 1,000 times. **Electron microscopes** are used to study detailed structures of a cell that cannot be easily seen or observed by light microscopy. They are capable of resolving structures as small as a few nanometers in length, such as individual virus particles or the pores on the surface of the nucleus.

## WHAT ARE THE DIFFERENT TYPES OF CELLS?

For centuries, scientists have known about cells. However, it wasn't until the development of the electron microscope that scientists were able to figure out what cells do. We now know that there are two distinct types of cells: **eukaryotic cells** and **prokaryotic cells**. A eukaryotic cell contains a membrane-bound structure called a nucleus and **cytoplasm**, filled with tiny structures called **organelles** (literally "little organs"). Examples of eukaryotic cells are *fungi*, *protists*, *plant cells*, and *animal cells*.

A prokaryotic cell, which is a lot smaller than a eukaryotic cell, lacks both a nucleus and membrane-bound organelles. Bacteria, for example, are prokaryotic cells. The genetic material in a prokaryote is one continuous, circular DNA molecule that lies free in the cell in an area called the **nucleoid**. In addition to a plasma membrane, most prokaryotes have a cell wall composed of peptidoglycan. Prokaryotes may also have ribosomes (although smaller than those found in eukaryotic cells) as well as one or more **flagella**, which are long projections used for motility (movement).

# Organelles

A eukaryotic cell is like a microscopic factory. It's filled with organelles, each of which has its own special tasks. Let's take a tour of a eukaryotic cell and focus on the structure and function of each organelle. Here's a picture of a typical animal cell and its principal organelles:

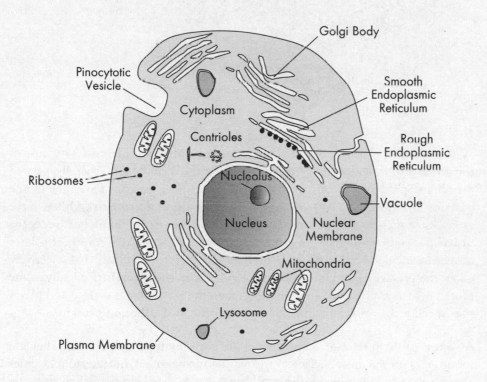

# Plasma Membrane

The cell has an outer envelope known as the **plasma membrane**. Although the plasma membrane appears to be a simple, thin layer surrounding the cell, it's actually a complex double-layered structure made up of phospholipids and proteins. The *hydrophobic* fatty acid tails face inward and the *hydrophilic* phosphate heads face outward.

The plasma membrane is important because it regulates the movement of substances into and out of the cell. The membrane itself is semipermeable, meaning that only certain substances, namely small hydrophobic molecules, pass through it unaided. Many proteins are associated with the cell membrane. Some of these proteins are loosely associated with the lipid bilayer (**peripheral proteins**). They are located on the inner or outer surface of the membrane. Others are firmly bound to the plasma membrane (**integral proteins**). These proteins are amphipathic, which means that their hydrophilic regions extend out of the cell or into the cytoplasm, while their hydrophobic regions interact with the tails of the membrane phospholipids. Some integral proteins do not extend all the way through the membrane (**transmembrane proteins**). This arrangement of phospholipids and proteins is known as the **fluid-mosaic model**.

proteins

Outside of cell

carbohydrate side chain

phospholipid bilayer

Inside of cell

Why should the plasma membrane need so many different proteins? It's because of the number of activities that take place in or on the membrane. Generally, plasma membrane proteins fall into several broad functional groups. Some membrane proteins form junctions between adjacent cells (**adhesion proteins**). Others serve as docking sites for proteins of the extracellular matrix or hormones (**receptor proteins**). Some proteins form pumps that use ATP to actively transport solutes across the membrane (**transport proteins**). Others form channels that selectively allow the passage of certain ions or molecules (**channel proteins**). Finally, some proteins, such as glycoproteins, are exposed on the extracellular surface and play a role in cell recognition and adhesion (**recognition and adhesion proteins**).

Attached to the surface of some proteins are **carbohydrate side chains.** They are found only on the outer surface of the plasma membrane. **Cholesterol** molecules are also found in the **phospholipid bilayer** because they help stabilize membrane fluidity in animal cells.

## The Nucleus

The nucleus, which is usually the largest organelle, is the control center of the cell. The nucleus not only directs what goes on in the cell, it is also responsible for the cell's ability to reproduce. It's the home of the hereditary information—DNA—which is organized into large structures called **chromosomes**. The most visible structure within the nucleus is the **nucleolus**, which is where rRNA is made and ribosomes are assembled.

## Ribosomes

The **ribosomes** are the sites of protein synthesis. Their job is to manufacture all the proteins required by the cell or secreted by the cell. Ribosomes are round structures composed of two subunits. The structure is composed of RNA and proteins. Ribosomes can be either free floating in the cell or attached to another structure called the **endoplasmic reticulum (ER)**.

## Endoplasmic Reticulum (ER)

The endoplasmic reticulum (ER) is a continuous channel that extends into many regions of the cytoplasm. The region of the ER that is "studded" with ribosomes is called the *rough ER* (RER). Proteins generated in the rough ER are trafficked to or across the plasma membrane. The region of the ER that lacks ribosomes is called the *smooth ER* (SER). The smooth ER makes lipids, hormones, and steroids and breaks down toxic chemicals.

## Golgi Bodies

The **Golgi bodies**, which look like stacks of flattened sacs, also participate in the processing of proteins. Once the ribosomes on the rough ER have completed synthesizing proteins, the Golgi bodies modify, process, and sort the products. They're the packaging and distribution centers for materials destined to be sent out of the cell. They package the final products in little sacs called **vesicles**, which carry the products to the plasma membrane. Golgi bodies are also involved in the production of lysosomes.

## Mitochondria: The Powerhouses of the Cell

Another important organelle is the *mitochondrion*. The **mitochondria** are often referred to as the "powerhouses" of the cell. They're power stations responsible for converting the energy from organic molecules into useful energy for the cell. The energy molecule in the cell is **adenosine triphosphate (ATP)**.

The mitochondrion is usually an easy organelle to recognize because it has a unique oblong shape and a characteristic double membrane consisting of an inner portion and an outer portion. The inner mitochondrial membrane forms folds known as *cristae*. As we'll see later, most of the production of ATP is done on the cristae.

**More Mitochondria = More Energy**

Since mitochondria are the cell's powerhouses, you're likely to find more of them in cells that require a lot of energy. Muscle cells, for example, are rich in mitochondria.

## Lysosomes

Throughout the cell are small, membrane-bound structures called **lysosomes**. These tiny sacs carry digestive enzymes, which they use to break down old, worn-out organelles, debris, or large ingested particles. The lysosomes make up the cell's cleanup crew, helping to keep the cytoplasm clear of unwanted flotsam. Lysosomes contain hydrolytic enzymes that only function at acidic pH, which is enclosed inside the lumen of the lysosome.

## Centrioles

The **centrioles** are small, paired, cylindrical structures that are found within **microtubule organizing centers (MTOCs)**. Centrioles are most active during cellular division. When a cell is ready to divide, the centrioles produce *microtubules*, which pull the replicated chromosomes apart and move them to opposite ends of the cell. Although centrioles are common in animal cells, they are not found in plant cells.

## Vacuoles

In Latin, the term *vacuole* means "empty cavity." But **vacuoles** are far from empty. They are fluid-filled sacs that store water, food, wastes, salts, or pigments.

## Peroxisomes

Peroxisomes are organelles that detoxify various substances, producing hydrogen peroxide as a byproduct. They also contain enzymes that break down hydrogen peroxide ($H_2O_2$) into oxygen and water. In animals, they are common in the liver and kidney cells.

## Cytoskeleton

Have you ever wondered what actually holds the cell together and enables it to keep its shape? The shape of a cell is determined by a network of fibers called the **cytoskeleton**. The most important fibers you'll need to know are **microtubules** and **microfilaments**.

Microtubules, which are made up of the protein **tubulin**, participate in cellular division and movement. These small fibers are an integral part of three structures: *centrioles*, *cilia*, and *flagella*. We've already mentioned that centrioles help chromosomes separate during cell division. **Cilia** and flagella are threadlike structures best known for their locomotive properties in single-celled organisms. The beating motion of cilia and flagella structures propels these organisms through their watery environments.

The two classic examples of organisms with these structures are the ***Euglena***, which gets about using its whiplike flagellum, and the ***Paramecium***, which is covered in cilia. The rhythmic beating of the *Paramecium*'s cilia enables it to motor about in waterways, ponds, and microscope slides in your biology lab. You've probably already checked these out in lab, but here's what they look like:

*Euglena* with its flagellum

*Paramecium* with its cilia

Though we usually associate such structures with microscopic organisms, they aren't the only ones with cilia and flagella. As you probably know, these structures are also found in certain human cells. For example, the cells lining your respiratory tract possess cilia that sweep constantly back and forth (beating up to 20 times per second), helping to keep dust and unwanted debris from descending into your lungs. And every sperm cell has a flagellum, which enables it to swim through the female reproductive organs to fertilize the waiting ovum.

Microfilaments, like microtubules, are important for movement. These thin, rod-like structures are composed of the protein actin, they are involved in cell mobility, and play a central role in muscle contraction.

## Plant Cells Versus Animal Cells

Plant cells contain most of the same organelles and structures seen in animal cells, with several key exceptions. Plant cells, unlike animal cells, have a protective outer covering called the **cell wall** (made of cellulose). A cell wall is a rigid layer just outside of the plasma membrane that provides support for the cell. It is found in plants, protists, fungi, and bacteria. (In fungi, the cell wall is usually made of **chitin**, a modified polysaccharide. Chitin is also a principle component of an arthropod's exoskeleton.) In addition, plant cells possess **chloroplasts** (organelles involved in photosynthesis). Chloroplasts contain chlorophyll, the light-capturing pigment that gives plants their characteristic green color. Another difference between plant and animal cells is that most of the cytoplasm within a plant cell is usually taken up by a large vacuole that crowds the other organelles. In mature plants, this vacuole contains the **cell sap**. Plant cells also differ from animal cells in that plant cells do not contain centrioles.

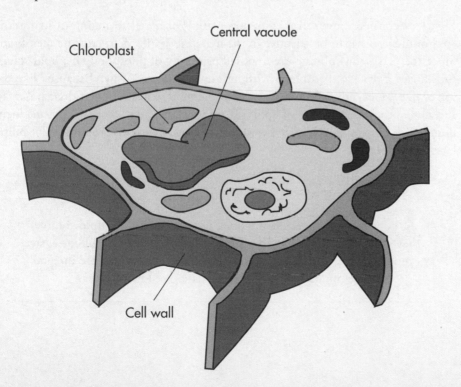

Central vacuole

Chloroplast

Cell wall

To help you remember the differences among prokaryotes, plant cells, and animal cells, we've put together this simple table. Make sure you learn it! ETS is bound to ask you which cells contain which structures:

| Structural Characteristics of Different Cell Types | | | |
|---|---|---|---|
| Structure | Prokaryote | Plant Cell | Animal Cell |
| Cell Wall | Yes | Yes | No |
| Plasma Membrane | Yes | Yes | Yes |
| Organelles | No | Yes | Yes |
| Nucleus | No | Yes | Yes |
| Centrioles | No | No | Yes |
| Ribosomes | Yes | Yes | Yes |

Why do we need to know about the structure of cells? Because biological structure is often closely related to function. (Watch out for this connection: It's a favorite theme for the AP Biology Exam.) And, more importantly, because ETS likes to test you on it!

# Transport: The Traffic Across Membranes

We've talked about the structure of cell membranes, now let's discuss how molecules and fluids pass through the plasma membrane. What are some of the patterns of membrane transport? The ability of molecules to move across the cell membrane depends on two things: (1) the semipermeability of the plasma membrane and (2) the size and charge of particles that want to get through.

First let's consider how cell membranes work. For a cell to maintain its internal environment, it has to be selective in the materials it allows to cross its membrane. Since the plasma membrane is composed primarily of phospholipids, lipid-soluble substances cross the membrane without any resistance. Why? Because "like dissolves like." Generally speaking, the lipid membrane has an open-door policy for substances that are made up of lipids. These substances can cross the plasma membrane without any problem. However, if a substance is hydrophilic, the bilipid layer won't let it in.

One exception to the rule is water:

> Although water molecules are polar (and therefore not lipid-soluble) they can rapidly cross a lipid bilayer (at a rate of about 3 billion water molecules a second, in fact) through aquaporins, which are integral membrane proteins that regulate the flow of water.

## Diffusion

We've just seen that lipid-soluble substances can traverse the plasma membrane without much difficulty. But what determines the *direction* of traffic across the membrane? Some substances move across a membrane by **simple diffusion**. That is, if there's a high concentration of a substance outside the cell and a low concentration inside the cell, the substance will move into the cell. In other words, the substance moves *down* a concentration gradient:

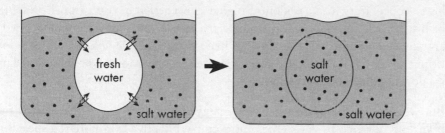

It's like riding a bicycle downhill. The bike "goes with the natural flow." Another name for this type of transport is **passive transport**. Here's one more thing you must remember: Simple diffusion does not require energy. A special kind of diffusion that involves the movement of water is called **osmosis**.

## Facilitated Transport

How do *lipid-insoluble* substances, which are dissolved in the fluid on either side of the cell membrane, get in and out of the cell? These dissolved substances, or **solutes**, rely on the proteins embedded in the plasma membrane. Special proteins—called **channel proteins**—can help lipid-insoluble substances get in or out:

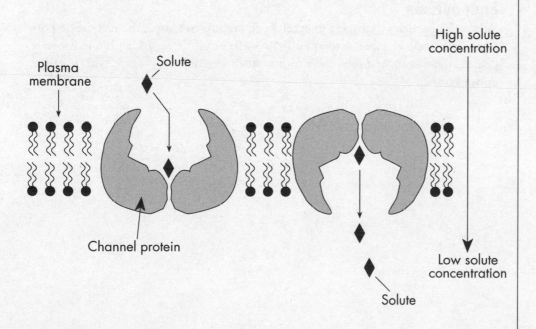

These proteins pick up the substance from one side of the membrane and carry it across to the other. This type of transport is known as **facilitated transport**, or facilitated diffusion. Facilitated diffusion is just like simple diffusion in one respect: The flow of the substance is *down* the concentration gradient. Therefore, it doesn't require any energy.

## Active Transport

Suppose a substance wants to move in the opposite direction—from a region of lower concentration to a region of higher concentration. A transport protein can help usher the substance across the plasma membrane, but it's going to need energy to accomplish this. This time it's like riding a bicycle uphill. Compared with riding downhill, riding uphill takes a lot more work. Movement against the natural flow is called **active transport**.

But where does the protein get this energy? Some proteins in the plasma membrane are powered by ATP. The best example of active transport is a special protein called the **sodium-potassium pump**. It ushers out sodium ions ($Na^+$) and brings in potassium ions ($K^+$) across the cell membrane. These pumps depend on ATP to get ions across that would otherwise remain in regions of higher concentration. Where do we usually find these proteins? In vertebrates, they're found in neurons and skeletal muscle fibers.

We've now seen that small substances can cross the cell membrane by:

- simple diffusion
- facilitated transport
- active transport

## Endocytosis

When the particles that want to enter a cell are just too large, the cell uses a portion of the cell membrane to engulf the substance. The cell membrane forms a pocket, pinches in, and eventually forms either a vacuole or a vesicle. This is called **endocytosis**.

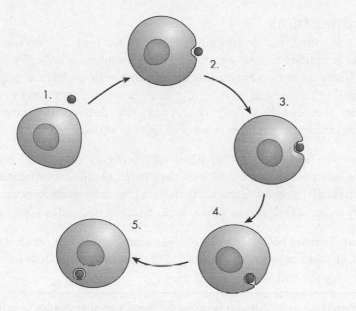

Three types of endocytosis exist: **pinocytosis**, **phagocytosis**, and **receptor-mediated endocytosis**. In pinocytosis, the cell ingests liquids ("cell-drinking"). In phagocytosis, the cell takes in solids ("cell-eating"). A special type of endocytosis, receptor-mediated endocytosis, involves cell surface receptors that are covered in clathrin-coated pits. (Clathrin is a kind of protein.) When a particle, or ligand, binds to one of these receptors, the ligand is brought into the cell by the invagination or "folding in" of the cell membrane. A vesicle then forms around the incoming ligand and carries it into the cell's interior.

## Bulk Flow

Other substances move by **bulk flow**. Bulk flow is the one-way movement of fluids brought about by *pressure*. For instance, the movement of blood through a blood vessel or movement of fluids in xylem and phloem of plants are examples of bulk flow.

## Dialysis

**Dialysis** is the diffusion of *solutes* across a selectively permeable membrane. For example, a cellophane bag is often used as an artificial membrane to separate small molecules from large molecules.

## Exocytosis

Sometimes large particles are transported *out* of the cell. In **exocytosis**, a cell ejects waste products or specific secretion products, such as hormones, by the fusion of a vesicle with the plasma membrane.

## Cell Junctions

When cells come in close contact with each other, they develop specialized **intercellular junctions** that involve their plasma membranes as well as other components. These structures may allow neighboring cells to form strong connections with each other, prevent passage of materials, or establish rapid communication between adjacent cells. There are three types of intercellular contact in animal cells: **desmosomes**, **gap junctions**, and **tight junctions**.

Desmosomes hold adjacent animal cells tightly to each other, like a rivet. They consist of a pair of discs associated with the plasma membrane of adjacent cells, plus the intercellular protein filaments that cross the small space between them. Intermediate filaments within the cells are also attached to the discs (see figure below).

Gap junctions are protein complexes that form channels in membranes and allow communication between the cytoplasm of adjacent animal cells or the transfer of small molecules and ions.

Tight junctions are tight connections between the membranes of adjacent animal cells. They're so tight that there is no space between the cells. Cells connected by tight junctions seal off body cavities and prevent leaks.

# KEY TERMS

cells
light microscopes
electron microscopes
eukaryotic cells
prokaryotic cells
cytoplasm
organelles
nucleoid
flagella
plasma membrane
peripheral proteins
integral proteins
transmembrane proteins
fluid-mosaic model
adhesion proteins
receptor proteins
transport proteins
channel proteins
recognition and adhesion proteins
carbohydrate side chains
cholesterol
phospholipid bilayer
chromosomes
nucleolus
ribosomes
endoplasmic reticulum (ER)
Golgi bodies
vesicles
mitochondria
adenosine triphosphate (ATP)
lysosomes
centrioles
microtubule organizing centers
   (MTOCs)
vacuoles

cytoskeleton
microtubules
microfilaments
tubulin
cilia
*Euglena*
*Paramecium*
cell wall
chitin
chloroplasts
cell sap
simple diffusion (or passive transport)
osmosis
solutes
channel proteins
facilitated transport (or facilitated
   diffusion)
active transport
sodium-potassium pump
endocytosis
pinocytosis
phagocytosis
receptor-mediated endocytosis
bulk flow
dialysis
exocytosis
intercellular junctions
desmosomes
gap junctions
tight junctions

# Summary

- Living things are composed of cells, the smallest unit of living organisms. Cells may be categorized into prokaryotes, which do not have a nucleus or membrane bound organelles, and eukaryotes, which have a nucleus and membrane-bound organelles. Features of prokaryotic cells (like bacteria, for example) include the following:

  - plasma membrane
  - flagella and cilia
  - cell wall
  - nucleoid region
  - ribosomes
  - bacterial chromosome
  - capsule
  - cytoplasm

- Components of the eukaryotic cells include the following:

  - nucleus
  - nucleolus
  - rough endoplasmic reticulum
  - smooth endoplasmic reticulum
  - Golgi body
  - ribosomes
  - mitochondria
  - vacuole
  - lysosome
  - peroxisomes
  - centrioles (animals only)
  - cytoplasm
  - plasma membrane
  - cytoskeleton (microtubules, microfilaments)
  - central vacuole (plants only)
  - chloroplasts (plants only)
  - cell wall (plants only)

- The plasma membrane is composed of a phospholipid bilayer. This membrane is hydrophobic at its core, which is responsible for the selective permeability of the membrane. Transport through the membrane is dependent on size and polarity of the molecules and concentration gradients. There are a few different types of cellular transport:

  - diffusion
  - facilitated transport
  - active transport
  - bulk flow
  - dialysis
  - exocytosis
  - endocytosis (pinocytosis, phagocytosis, receptor-mediated endocytosis)

- Cells are connected by tight junctions, desmosomes, and gap junctions, which all function to connect adjacent cells.

# Chapter 5 Drill

Answers and explanations can be found in Chapter 16.

1. Movement of substances into the cell is largely dependent on the size, polarity, and concentration gradient of the substance. Which of the following represents an example of active transport of a substance into a cell?

   (A) Diffusion of oxygen into erythrocytes (red blood cells) in the alveolar capillaries of the lungs
   (B) Influx of sodium ions through a voltage-gated ion channel in a neuron cell during an action potential
   (C) The sodium-potassium pump, which restores resting membrane potentials in neurons through the use of ATP
   (D) Osmosis of water into an epithelial cell lining the lumen of the small intestine

2. The development of electron microscopy has provided key insights into many aspects of cellular structure and function which had previously been too small to be seen. All of the following would require the use of electron microscopy for visualization EXCEPT

   (A) the structure of a bacteriophage
   (B) the matrix structure of a mitochondrion
   (C) the shape and arrangement of bacterial cells
   (D) the pores on the nuclear membrane

3. *Vibrio cholerae* (shown below) are highly pathogenic bacteria that are associated with severe gastrointestinal illness and are the causative agent of cholera. In extreme cases, antibiotics are prescribed that target bacterial structures that are absent in animal cells. Which of the following structures is most likely targeted by antibiotic treatment?

   (A) Flagella
   (B) Plasma membrane
   (C) Ribosomes
   (D) Cell wall

Question 4 is based on the information and table below.

A new unicellular organism has recently been identified living in thermal pools in Yellowstone National Park. The thermal pools have average temperatures of 45°C, a pH of approximately 3.2, and have high concentrations of sulfur-containing compounds. To identify the organism, a microbiologist performs a series of tests to evaluate its structural organization. The table below summarizes the microscopy data of the newly identified organism.

| Cellular Structure | Analysis |
|---|---|
| Plasma Membrane | Present |
| Cell Wall | Present, Very Thick |
| Mitochondria | Absent |
| Ribosomes | Present, Highly Abundant |
| Flagella | Present, Peritrichous Organization |

4. This organism is most likely a new species of which of the following?

   (A) Algae
   (B) Protozoa
   (C) Bacteria
   (D) Fungi

Question 5 represents a question requiring a numeric answer.

5. Chlorophyll is a green pigment present in the chloroplasts of algae and plants. It is essential for catalyzing the light-dependent cycles in photosynthesis. A scientist purifies both forms of chlorophyll (*a* and *b*) from plant chloroplasts and evaluates them for light absorption using a spectrophotometer. Using the spectrophotometer data provided above, at what wavelength is the absorbance of chlorophyll *a* at its maximum? Give your answer to the nearest whole number.

6. In a eukaryotic cell, which of the following organelles work together?

   (A) Nuclear envelope, nucleolus, vacuoles, centrioles
   (B) Ribosomes, rough endoplasmic reticulum, Golgi bodies, plasma membrane
   (C) Mitochondria, ribosomes, lysosomes, chloroplasts
   (D) Centrioles, nucleolus, smooth endoplasmic reticulum, lysosomes

7. A lipid-soluble hormone, estrogen, is secreted from the ovaries. This molecule travels through the body via the bloodstream. A researcher was interested in reducing estrogen's effect in order to determine the response of decreased estrogen on the organism. Which of the following is a valid strategy for reducing effects of estrogen on the whole research organism?

   (A) Treat with a competitive inhibitor drug that blocks all receptors at the plasma membrane
   (B) Treat with lipid-soluble testosterone
   (C) Treat with a lipid-soluble noncompetitive inhibitor that specifically reduces estrogen binding to the intracellular receptor
   (D) Remove ovaries of the organism

8. Which type of cell would be the most useful for studying the endomembrane system?

   (A) Neurons
   (B) Pancreatic cells
   (C) *E. coli*
   (D) Leukocytes

9. *Paramecium* is a single-celled protist that lives in freshwater habitats. In these conditions, *Paramecium* has evolved strategies to handle the potential consequences of inhabiting this hypotonic environment. One of these strategies could be

   (A) contractile vacuoles, which expel water forcefully
   (B) increased aquaporins in its cellular membrane
   (C) many cilia covering its surface
   (D) salt receptors on its surface to seek out highly concentrated areas

10. The $Na^+K^+$ sodium potassium pump transports sodium and potassium ions across the plasma membrane against their concentration gradients. This pump is not considered a co-transporter because

    (A) ATP is produced through this transporter
    (B) a gradient, not ATP, drives the transport of these ions
    (C) ATP, not a gradient, drives the transport of these ions
    (D) cotransporters go in only one direction, but the sodium-potassium pump drives ions across the membrane in both directions

# REFLECT

Respond to the following questions:

- For what content topics discussed in this chapter do you feel you have achieved sufficient mastery to answer multiple-choice questions correctly?

- For which content topics discussed in this chapter do you feel you have achieved sufficient mastery to discuss effectively in a free response?

- For which content topics discussed in this chapter do you feel you need more work before you can answer multiple-choice questions correctly?

- For which content topics discussed in this chapter do you feel you need more work before you can discuss effectively in a free response?

- What parts of this chapter are you going to re-review?

- Will you seek further help, outside this book (such as a teacher, tutor, or AP Central), on any of the content in this chapter—and if so, on what content?

# Chapter 6
# Cellular Energetics

# BIOENERGETICS

In Chapter 4, we discussed some of the more important organic molecules. But what makes these molecules so important? Glucose, starch, and fat are all energy-rich. However, the energy is packed in the chemical bonds holding the molecules together. To carry out the processes necessary for life, cells must find a way to release the energy in these bonds when they need it and store it away when they don't. The study of how cells accomplish this is called **bioenergetics**. Generally, bioenergetics is the study of how energy from the sun is transformed into energy in living things.

All the energy for life comes from chemical bonds. During chemical reactions, such bonds are either broken or formed. This process involves energy, no matter in which direction we go. Every chemical reaction involves a change in energy.

Energy is invested in the formation of bonds, whereas energy is released when bonds are broken. However, the breaking apart of chemical bonds requires the input of energy. The way that cells are able to break these bonds without spending large amounts of energy is through the use of special molecules called **enzymes**.

# ENZYMES

Most chemical reactions do not occur haphazardly in the cell. To help control the chemical reactions essential for life, cells rely on enzymes. These proteins help "kick-start" reactions and speed them up once they get rolling, enabling cells to get the most out of the energy sources available to them. Enzymes are **organic catalysts**; they speed up the rate of a reaction without altering the reaction itself.

That is, they *catalyze* them without being changed in the reaction themselves. Before we discuss precisely how this works, let's review some of the different types of reactions that can occur within a cell.

# TYPES OF REACTIONS

**Exergonic reactions** are those in which the products have *less* energy than the reactants. Simply put, energy is given off during the reaction.

Let's look at an example. The course of a reaction can be represented by an energy diagram. Here's an energy diagram for an exergonic reaction:

You'll notice that energy is represented along the *y*-axis. Based on the diagram, our reaction released energy. An example of an exergonic reaction is when food is oxidized in mitochondria of cells and then releases the energy stored in the chemical bonds.

Reactions that require an input of energy are called **endergonic reactions**. You'll notice that the products have *more* energy than the reactants.

The products gained energy in the form of heat. An example is when plants use carbon dioxide and water to form sugars.

## Activation Energy

Although we said that exergonic reactions release energy, this does not mean that they do not require any energy to get started. Take a look at this energy diagram of a typical exergonic reaction:

You'll notice that we needed a little energy to get us going. That's because chemical bonds must be broken before new bonds can form. This energy barrier—the hump in the graph—is called the **activation energy**. Once a set of reactants has reached its activation energy, the reaction can occur much faster than it would in the absence of the enzyme.

## Getting Back to Enzymes

Why are enzymes so important in biology? They're important because many reactions would never occur in the cell if it weren't for the help of enzymes. As we saw earlier, enzymes, by definition, catalyze reactions: They activate them. What this means in chemical terms is that they *lower* the activation energy of a reaction, enabling the reaction to occur much faster than it would in the absence of the enzyme.

## Enzyme Specificity

Most of the crucial reactions that occur in the cell require enzymes. Yet enzymes themselves are highly specific—in fact, each enzyme catalyzes only one kind of reaction. This is known as **enzyme specificity**. Because of this, enzymes are usually named after the molecules they target. In enzymatic reactions, the targeted molecules are known as **substrates**. For example, maltose, a disaccharide, can be broken down into two glucose molecules. Our substrate, maltose, gives its name to the enzyme that catalyzes this reaction: *maltase*.

Many enzymes are named simply by replacing the suffix of the substrate with *–ase*. Using this nomenclature, malt*ose* becomes malt*ase*.

## Enzyme-Substrate Complex

Enzymes have a unique way of helping reactions along. As we just saw, the reactants in an enzyme-assisted reaction are known as substrates. During a reaction, the enzyme's job is to bring the substrates together. It accomplishes this through a special region on the enzyme known as an **active site**.

The enzyme temporarily binds the substrates to its active site and forms an **enzyme-substrate complex**. Let's take a look:

Once the reaction has occurred and the product is formed, the enzyme is released from the complex and restored to its original state. Now the enzyme is free to react again with another bunch of substrates.

By binding and releasing over and over again, the enzyme speeds the reaction along, enabling the cell to release much-needed energy from various molecules. Here is a quick review on the function of enzymes.

**Enzymes Do**

- increase the rate of a reaction by lowering the reaction's activation energy
- form temporary enzyme-substrate complexes
- remain unaffected by the reaction

**Enzymes Don't**

- change the reaction
- make reactions occur that would otherwise not occur at all

## Induced Fit

However, scientists have discovered that enzymes and substrates don't fit together quite so seamlessly. It appears that the enzyme has to change its shape slightly to accommodate the shape of the substrates. This is called **induced fit**.

## Enzymes Don't Always Work Alone

Enzymes sometimes need a little help in catalyzing a reaction. Those factors are known as **coenzymes**. Vitamins are examples of organic coenzymes. Your daily dose of vitamins is important for just this reason: The vitamins are "active and necessary participants" in crucial chemical reactions. The function of coenzymes is to accept electrons and pass them along to another substrate. Two examples of such enzymes are $NAD^+$ and $NADP^+$.

In addition to organic coenzymes, inorganic elements—called **cofactors**—help catalyze reactions. These elements are usually metal ions, such as $Fe^{+2}$, which is a key cofactor of hemoglobin because iron is important for binding oxygen in your bloodstream.

## Factors Affecting Reaction Rates

Enzymatic reactions can be influenced by a number of factors, such as temperature, pH, and the relative amounts of enzyme and substrate.

### Temperature

The rate of a reaction increases with increasing temperature, up to a point, because an increase in the temperature of a reaction increases the chance of collisions among the molecules. But too much heat can damage an enzyme. If a reaction is conducted at an excessively high temperature (above 42°C), the enzyme loses its three-dimensional shape and becomes inactivated. Enzymes damaged by heat and deprived of their ability to catalyze reactions are said to be *denatured*.

Here's one thing to remember: All enzymes operate at an ideal temperature. For most human enzymes, this temperature is body temperature, 37°C.

## pH

Enzymes also function best at a particular pH. For most enzymes, the optimal pH is at or near a pH of 7:

Other enzymes operate at a low pH. For instance, pepsin, the digestive enzyme found in the stomach, is most effective at an extremely acidic pH of 2.

Here's another important piece of information: Most enzymes are active only over a narrow range of pH.

## Enzyme Regulation

We know that enzymes control the rates of chemical reactions. But what regulates the activity of enzymes? It turns out that a cell can control enzymatic activity by regulating the conditions that influence the shape of the enzyme. For example, some enzymes have **allosteric sites**, a region of the enzyme other than the active site to which a substance can bind. Substances called **allosteric regulators** can either inhibit or activate enzymes. An **allosteric inhibitor** will bind to an allosteric site and keep the enzyme in its inactive form while an **allosteric activator** will bind to an enzyme and induce its active form. Allosteric enzymes are subject to **feedback inhibition** in which the formation of an end product inhibits an earlier reaction in the sequence.

Most enzymes can be inhibited by certain chemical substances. If the substance has a shape that fits the active site of an enzyme, it can compete with the substrate and effectively inactivate the enzyme. This is called **competitive inhibition**. Usually, a competitive inhibitor is structurally similar to the normal substrate. In **noncompetitive inhibition**, the inhibitor binds with the enzyme at a site other than the active site and inactivates the enzyme by altering its shape. This prevents the enzyme from binding with the substrate at the active site.

# ADENOSINE TRIPHOSPHATE (ATP)

We've all heard the expression *nothing is for free*. The same holds true in nature. Here's a fundamental principle of energy that it is necessary to address:

> Energy cannot be created or destroyed. In other words, the sum of energy in the universe is constant.

This rule is called the **first law of thermodynamics**. As a result, the cell cannot take energy out of thin air. Rather, it must harvest it somewhere.

The **second law of thermodynamics** states that energy transfer leads to less organization. That means the universe tends toward disorder (or **entropy**).

As we just saw, almost everything an organism does requires energy. How, then, can the cell acquire the energy it needs without becoming a major mess? Fortunately, it's through adenosine triphosphate (ATP).

ATP, as the name indicates, consists of a molecule of adenosine bonded to three phosphates. The great thing about ATP is that an enormous amount of energy is packed into those phosphate bonds, particularly the third bond.

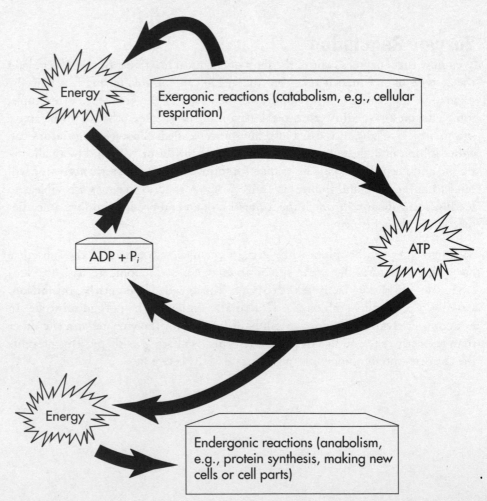

When a cell needs energy, it takes one of these potential-packed molecules of ATP and splits off the third phosphate, forming adenosine diphosphate (ADP) and one loose phosphate ($P_i$), while releasing energy in the process:

$$ATP \rightarrow ADP + P_i + energy$$

The energy released from this reaction can then be put to whatever use the cell so pleases. Of course, this doesn't mean that the cell is above the laws of thermodynamics. But within those constraints, ATP is the best source of energy the cell has available. It is relatively neat (only one bond needs to be broken to release that energy) and relatively easy to form. Organisms can use exergonic processes that increase energy, like breaking down ATP, and to power endergonic reactions, like building organic macromolecules.

## Sources of ATP

But where does all this ATP come from? It is produced in one of two ways: (1) through photosynthesis, or (2) through cellular respiration.

**Photosynthesis** involves the transformation of solar energy into chemical energy. Plants take carbon dioxide, water, and energy (in the form of sunlight) and use them to produce glucose. The overall reaction is:

$$6CO_2 + 6H_2O + sunlight \rightarrow C_6H_{12}O_6 + 6O_2$$

We'll hold off on a full discussion of photosynthesis until Chapter 7. For now, let's focus on the other means of ATP production—**cellular respiration**.

## Cellular Respiration: The Shorthand Version

In cellular respiration, which is performed by all organisms, ATP is produced through the breakdown of nutrients. You'll recall from the beginning of this chapter that many organic molecules are important to cells because they are energy-rich. This is where that energy comes into play.

In the shorthand version, cellular respiration looks something like this:

$$C_6H_{12}O_6 + 6O_2 \rightarrow 6CO_2 + 6H_2O + ATP$$

Notice that we've taken a sugar, perhaps a molecule of glucose, and combined it with oxygen and water to produce carbon dioxide, water, and energy in the form of our old friend, ATP. However, as you probably already know, the actual picture of what really happens is far more complicated.

Generally speaking, we can break cellular respiration down to two different approaches: aerobic respiration or anaerobic respiration. If ATP is made in the presence of oxygen, we call it **aerobic respiration**. If oxygen isn't present, we call it **anaerobic respiration**. Let's jump right in with aerobic respiration.

## AEROBIC RESPIRATION

Aerobic respiration consists of four stages:

1. Glycolysis
2. Formation of acetyl CoA
3. The Krebs (or citric acid) cycle
4. Oxidative phosphorylation

There are so many steps within each stage that some students find this topic too confusing to follow. Don't sweat it. We've come up with a simple method to keep all the stages of cellular respiration in order: Just keep track of the number of carbons at each stage.

## Stage 1: Glycolysis

The first stage begins with **glycolysis**, the splitting (-*lysis*) of glucose (*glyco-*). Glucose is a six-carbon molecule that is broken into two three-carbon molecules called **pyruvic acid**. This breakdown of glucose also results in the net production of two molecules of ATP:

Glucose + 2 ATP + 2 NAD⁺ → 2 Pyruvic acid + 4 ATP + 2 NADH

Although we've written glycolysis as if it were a single reaction, this process doesn't occur in one step. In fact, it requires a sequence of enzyme-catalyzed reactions!

Fortunately, you don't need to memorize these steps for the test. What you do need to know is that glucose doesn't *automatically* generate ATP. It has to be activated. Once glucose is phosphorylated, it eventually splits into pyruvate.

If you take a good look at the reaction above, you'll see two ATPs are needed to produce four ATPs. You've probably heard the expression, "You have to spend money to make money." In biology, you have to invest ATP to make ATP: Our investment of two ATPs yielded four ATPs, for a net gain of two. The ATP molecules generated in glycolysis are phosphorylated by substrate-level phosphorylation, with the direct bonding of ADP to inorganic phosphate with the help of of an enzyme.

A second product in glycolysis is 2 NADH, which results from the transfer of $H^+$ to the hydrogen carrier $NAD^+$. NADH will be used elsewhere in respiration to make additional ATP.

There are four important tidbits to remember regarding glycolysis:

- occurs in the cytoplasm
- net of 2 ATPs produced
- 2 pyruvic acids formed
- 2 NADH produced

Once the cell has undergone glycolysis, it has two options: It can continue anaerobically, or it can switch to true aerobic respiration. As we'll soon see, the cell's decision has a lot to do with the environment in which it finds itself. If oxygen is present, many cells switch directly to aerobic respiration. If no oxygen is present, those same cells may carry out anaerobic respiration. Still others have no choice, and carry out only anaerobic respiration, with or without oxygen.

Since ETS is more likely to ask you about aerobic respiration, we'll look closely at the remaining steps.

However, before we do so, let's jump back to those important organelles, the mitochondria. We already know from our discussion in Chapter 5 that the mitochondria are the sites of cellular respiration. Now it's time to see exactly *where* they manufacture ATP.

The double membrane of the mitochondria divides the organelle into four regions:

- **matrix**
- **inner mitochondrial membrane**
- **intermembrane space**
- **outer membrane**

Let's have a look:

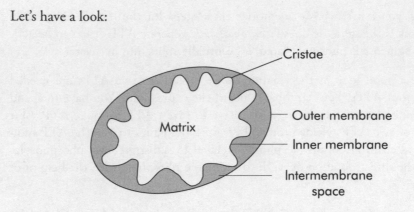

Why do you need to know about the different regions within a mitochondrion? Because we'll soon see that several of the stages of aerobic respiration occur within these regions of the mitochondria—and ETS loves to ask you questions about where things occur! Keep them in mind. We'll be discussing them on the following pages.

## Stage 2: Formation of Acetyl CoA

When oxygen is present, pyruvic acid is transported to the mitochondrion. Each pyruvic acid (a three-carbon molecule) is converted to **acetyl coenzyme A** (a two-carbon molecule) and $CO_2$ is released:

2 Pyruvic acid + 2 Coenzyme A + 2 NAD$^+$ → 2 Acetyl CoA+ 2 $CO_2$ + 2 NADH

Are you keeping track of our carbons? We've now gone from two three-carbon molecules to two two-carbon molecules. The extra carbons leave the cell in the form of $CO_2$. Once again, two molecules of NADH are also produced.

## Stage 3: The Krebs Cycle

The next stage is the **Krebs cycle**, also known as the **citric acid cycle**. Each of the two acetyl coenzyme A molecules will enter the Krebs cycle, one at a time, and all the carbons will ultimately be converted to $CO_2$. This stage occurs in the matrix of the mitochondria.

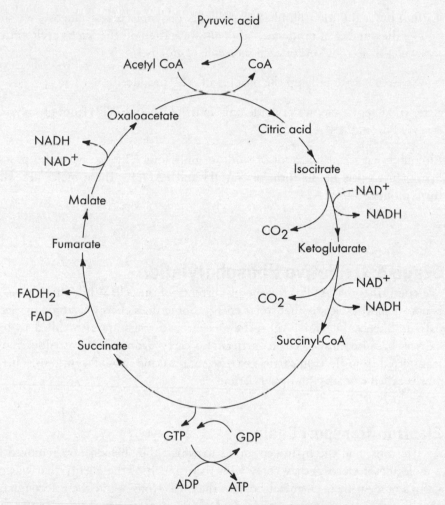

Let's track the carbons again. Each molecule of acetyl CoA produced from the second stage of aerobic respiration combines with **oxaloacetate**, a four-carbon molecule, to form a six-carbon molecule, **citric acid** or citrate:

Since the cycle begins with a four-carbon molecule, oxaloacetate, it also has to end with a four-carbon molecule to maintain the cycle. So how many carbons do we have to lose to keep the cycle going? Two carbons, both of which will be released as $CO_2$. Now the cycle is ready for another turn with the second acetyl CoA.

With each turn of the cycle, three additional types of molecules are produced:

- 1 ATP
- 3 NADH
- 1 FADH$_2$

To figure out the total number of products per molecule of glucose, we simply double the number of products—after all, we started off the Krebs cycle with two molecules of acetyl CoA for each molecule of glucose!

Now we're ready to tally up the number of ATP produced.

After the Krebs cycle, we've made only four ATP—two ATP from glycolysis and two ATP from the Krebs cycle.

Although that seems like a lot of work for only four ATP, we have also produced hydrogen carriers in the form of NADH and FADH$_2$. These molecules will, in turn, produce lots of ATP.

## Stage 4: Oxidative Phosphorylation

We said earlier that ATP is the energy currency of the cell. While this is true, ATP is not the only molecule that stores energy. Sometimes energy is stored by electron carriers like NAD$^+$ and FAD. (These electron carriers are also called hydrogen carriers because most electron carriers also carry hydrogen atoms.) Electrons are transferred from electron carriers to oxygen, resulting in ATP synthesis. This process is called oxidative phosphorylation.

### Electron Transport Chain

As electrons (and the hydrogen atoms to which they belong) are removed from a molecule of glucose, they carry with them much of the energy that was originally stored in their chemical bonds. These electrons—and their accompanying energy—are then transferred to readied hydrogen carrier molecules. In the case of cellular respiration, these charged carriers are NADH and FADH$_2$.

Let's see how many "loaded" electron carriers we've produced. We now have:

- 2 NADH molecules from glycolysis
- 2 NADH from the production of acetyl CoA
- 6 NADH from the Krebs cycle
- 2 FADH$_2$ from the Krebs cycle

That gives us 12 altogether.

Now let's consider the fate of all the electrons removed from the breakdown of glucose. Here's what happens. The electron carriers—NADH and FADH₂—"shuttle" electrons to the electron transport chain, and the hydrogen atoms are split into hydrogen ions and electrons:

$$H_2 \rightarrow 2\ H^+ + 2e^-$$

Then, two interesting things occur. First, the high-energy electrons from NADH and FADH₂ are passed down the electron transport chain, which is a series of protein carrier molecules that are embedded in the cristae, the membrane along the inner membrane of a mitochondrion. Some of the carrier molecules in the electron transport chain are iron-containing carriers called **cytochromes**. Take a look at the entire chain below.

Each carrier molecule hands down the electrons to the next molecule in the chain. The electrons travel down the electron transport chain until they reach the final electron acceptor, oxygen. Oxygen combines with these electrons (and some hydrogens) to form water. This explains the "aerobic" in aerobic respiration. If oxygen weren't available to accept the electrons, they wouldn't move down the chain at all, thereby shutting down the whole process of ATP production.

## CHEMIOSMOSIS

At the same time that electrons are being passed down the electron transport chain, another mechanism is at work. Remember those hydrogen ions (also called *protons*) that split off from the original hydrogen atom? Some of the energy released from the electron transport chain is used to pump hydrogen ions across the inner mitochondrial membrane to the intermembrane space. The pumping of hydrogen ions into the intermembrane space creates a **pH gradient**, or **proton gradient**. The potential energy established in this gradient is responsible for the production of ATP.

## Mitochondrial Chemiosmosis

These hydrogen ions can diffuse across the inner membrane only by passing through channels called **ATP synthase**. Meanwhile, ADP and $P_i$ are on the other side of these protein channels. The flow of protons through these channels produces ATP by combining ADP and $P_i$ on the matrix side of the channel. This process is called **oxidative phosphorylation**.

Three other things you're expected to know for the AP Biology Exam:

- Every NADH yields 3 ATP (except NADH from glycolysis produces 2 ATP).
- Every $FADH_2$ yields 2 ATP.
- The total number of ATP produced in Stage 4 is 32 ATP.

Don't concern yourself with the exact number of NADH, $FADH_2$, $CO_2$, and ATP produced at each stage.

For the AP Biology Exam, all you need to remember are the big steps and the overall outcome:

| STAGES OF AEROBIC RESPIRATION | | | |
|---|---|---|---|
| Process | Location | Input | Output |
| Glycolysis | Cytoplasm | 1 Glucose<br>2 ATP<br>2 NAD$^+$ | 2 pyruvates<br>**4 ATP**<br>2 NADH |
| Formation of Acetyl-CoA | As pyruvate is transported into the mitochondria | 2 pyruvates<br>2 NAD$^+$<br>2 Coenzyme A | 2 Acetyl-CoA<br>2 NADH<br>2 CO$_2$ |
| Krebs or Citric Acid Cycle | Matrix of mitochondria | 2 Acetyl-CoA<br>4 NAD$^+$<br>2 FADH | 6 NADH<br>2 FADH$_2$<br>**2 ATP** |
| Oxidative Phosphorylation and Chemiosmosis | Inner mitochondrial membrane | 10 NADH<br>2 FADH$_2$<br>2 O$_2$ | 2 H$_2$O<br>**32 ATP** |
| | | | **Overall Net:**<br>**36 ATP** |

# ANAEROBIC RESPIRATION

Some organisms can't undergo aerobic respiration. They're anaerobic. They can't use oxygen to make ATP. We just learned that oxygen is important because it's the final electron acceptor in the electron transport chain.

How do anaerobic organisms derive energy? Since glycolysis is an anaerobic process, they can make 2 ATP from this stage. However, instead of carrying out the other stages of aerobic respiration (the Krebs cycle, electron transport chain, and oxidative phosphorylation), these organisms carry out a process called **fermentation**. Under anaerobic conditions, pyruvic acid is converted to either **lactic acid** or **ethyl alcohol** (or **ethanol**) and carbon dioxide.

For the AP Biology Exam, you should remember the two pathways anaerobic organisms undergo: glycolysis and fermentation. Unfortunately, anaerobic respiration is not very efficient. It only results in a gain of 2 ATP for each molecule of glucose broken down.

As you can see from the chart below, there are two basic end products in anaerobic respiration. In both pathways, the NADH formed during glycolysis reduces (adds hydrogen to) pyruvate.

What types of organisms undergo fermentation?

- Yeast cells and some bacteria make ethanol and carbon dioxide.
- Other bacteria produce lactic acid.

## Your Muscle Cells Can Ferment

Although human beings are aerobic organisms, they can actually carry out fermentation in their muscle cells. Have you ever had a cramp? If so, that cramp was the consequence of anaerobic respiration.

When you exercise, your muscles require a lot of energy. To get this energy, they convert enormous amounts of glucose to ATP. But as you continue to exercise, your body doesn't get enough oxygen to keep up with the demand in your muscles. This creates an oxygen debt. What do your muscle cells do? They switch over to anaerobic respiration. Pyruvic acid produced from glycolysis is converted to lactic acid. As a consequence, the lactic acid causes your muscles to ache.

# KEY TERMS

bioenergetics

enzymes

organic catalysts

exergonic reactions

endergonic reactions

activation energy

enzyme specificity

substrates

active site

enzyme-substrate complex

induced fit

coenzymes

cofactors

allosteric sites

allosteric regulators

allosteric inhibitor

allosteric activator

feedback inhibition

competitive inhibition

noncompetitive inhibition

first law of thermodynamics

second law of thermodynamics

entropy

photosynthesis

cellular respiration

aerobic respiration

anaerobic respiration

glycolysis

pyruvic acid

matrix

inner mitochondrial membrane

intermembrane space

outer membrane

acetyl coenzyme A

Krebs cycle (or citric acid cycle)

oxaloacetate

citric acid

cytochromes

pH gradient
   (or proton gradient)

ATP synthase

oxidative phosphorylation

fermentation

lactic acid

ethyl alcohol (or ethanol)

# Summary

- The study of how cells accomplish biological processes is called bioenergetics.

    ○ Chemical reactions are catalyzed by enzymes.
    ○ Enzymes lower the activation energy of chemical reactions.
    ○ Enzymes do not change the energy difference between reactants and products, but by lowering the activation energy, they facilitate chemical reactions to occur.

- Enzymes are proteins that are highly specific for their substrate (reactant).

- Enzymes have an optimal, narrow range of temperature and pH in which they have the highest rate of reaction. Outside this range, they may undergo denaturation and, therefore, no longer be active.

- Enzyme activity may also be regulated or altered by allosteric regulators, inhibitors, and activators.

- Enzymes may require coenzymes or cofactors to help catalyze reactions.

- Cellular respiration consumes oxygen and produces a lot of ATP energy for the cell, using NADH and $FADH_2$ as electron carriers. There are four stages to cellular respiration:

    ○ glycolysis
    ○ formation of acetyl-CoA
    ○ Krebs cycle
    ○ oxidative phosphorylation and chemiosmosis

- Anaerobic respiration occurs when cells lack oxygen to act as a final electron acceptor. In this process, the pyruvates generated by glycolysis are broken down by fermentation to produce lactic acid or ethanol and $NAD^+$ and ATP, which permits glycolysis to continue.

# Chapter 6 Drill

Answers and explanations can be found in Chapter 16.

1. The mitochondrion is a critical organelle structure involved in cellular respiration. Below is a simple schematic of the structure of a mitochondrion. Which of the structural components labeled below in the mitochondrion is the primary location of the Krebs cycle?

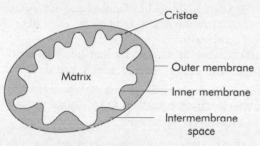

(A) Inner membrane
(B) Matrix
(C) Intermembrane space
(D) Outer membrane

2. Binding of Inhibitor Y as shown below inhibits a key catalytic enzyme by inducing a structural conformation change. Which of the following accurately describes the role of Inhibitor Y?

(A) Inhibitor Y competes with substrates for binding in the active site and functions as a competitive inhibitor.
(B) Inhibitor Y binds allosterically and functions as a competitive inhibitor.
(C) Inhibitor Y competes with substrates for binding in the active site and functions as a non-competitive inhibitor.
(D) Inhibitor Y binds allosterically and functions as a non-competitive inhibitor

3. A single step chemical reaction is catalyzed by the addition of an enzyme. Which of the reaction coordinate diagrams accurately shows the effect of the added enzyme (represented by the dashed line) to the reaction?

Questions 4–6 refer to the following diagram and paragraph.

Glycolysis (shown below) is a critical metabolic pathway that is utilized by nearly all forms of life. The process of glycolysis occurs in the cytoplasm of the cell and converts 1 molecule of glucose into 2 molecules of pyruvic acid.

4. How many net ATP would be generated directly from glycolysis from the breakdown of 2 glucose molecules?

(A) 2
(B) 4
(C) 8
(D) 12

5. Glycolysis does not require oxygen to occur in cells. However, under anaerobic conditions, glycolysis normally requires fermentation pathways to occur to continue to produce ATP. Which best describes why glycolysis is dependent on fermentation under anaerobic conditions?

(A) Glycolysis requires fermentation to produce more glucose as a substrate.
(B) Glycolysis requires fermentation to synthesize lactic acid that is used as an enzyme cofactor of hexokinase (which catalyzes the first step of glycolysis).
(C) Glycolysis requires fermentation to generate ATP molecules to complete the early steps of the pathway.
(D) Glycolysis requires fermentation to generate pyruvate for a later step in the pathway.

6. Which of the following most accurately describes the net reaction of glycolysis?

(A) It is an endergonic process because it results in a net increase in energy.
(B) It is an exergonic process because it results in a net increase in energy.
(C) It is an endergonic process because it results in a net decrease in energy.
(D) It is an exergonic process because it results in a net decrease in energy.

7. *Taq* polymerase, a DNA polymerase derived from thermophilic bacteria, is used in polymerase chain reactions (PCR) in the laboratory. During PCR, *Taq* catalyzes DNA polymerization, similar to how it would in bacteria. A normal PCR cycle is as follows:

| | |
|---|---|
| 1. Melting/Denaturing | 95°C |
| 2. Primer Annealing | 50°C |
| 3. Elongation of DNA | 72°C |
| (repeat 20–30 cycles) | |

Which of the following conditions likely describes the living environment of *Taq* bacteria?

(A) Freshwater with acidic pH
(B) Hydrothermal vents reaching temperatures between 70–75°C
(C) Hot springs of 40°C
(D) Tide pools with high salinity

8. The biggest difference between an enzyme-catalyzed reaction and an uncatalyzed reaction is

   (A) the free energy between the reactants and the products
   (B) the free energy difference between the reactants and the products does not change
   (C) the catalyzed reaction would not occur without the enzyme
   (D) the uncatalyzed reaction would be unlikely to occur in a biologically relevant timeframe without the enzyme.

9. Two groups of cells were grown under identical conditions. Mitochondria from each group was isolated and placed in a low pH (approximately pH 6.8) condition. Both samples were exposed to oxygen bubbles through the growth media. What would you expect to see in terms of ATP production in the sample of cells placed in a low pH, with respect to the control population?

   (A) ATP production decreases.
   (B) ATP production increases.
   (C) ATP production stays the same.
   (D) ATP initially increases, then decreases.

10. Rotenone acts as a common and potent poison that works by inhibiting complex I of the electron transport chain. In cells treated with rotenone, NADH is no longer capable of accepting and passing along electrons from complex I to ubiquinone. A researcher treats a group of cells with rotenone and measures oxygen consumption over time. Oxygen consumption decreases, but does not cease. How would you explain this result?

    (A) The concentration of rotenone is not high enough; the inhibition is not effective.
    (B) $FADH_2$ donates electrons to complex II, bypassing the inhibited complex I.
    (C) Oxygen is consumed elsewhere in the cells.
    (D) Oxygen is consumed by complex IV, which is not inhibited in this case.

11. The second law of thermodynamics states that entropy, or disorder, is constantly increasing in the universe for spontaneous processes. Therefore, how is it possible that organisms exist in very ordered states?

    (A) The second law of thermodynamics does not apply to biological life.
    (B) Energy can be created or destroyed.
    (C) The catabolic reactions in metabolism balance the effect of the anabolic reactions.
    (D) Biological life creates an increase in entropy through dissemination of heat and waste.

## REFLECT

Respond to the following questions:

- For what content topics discussed in this chapter do you feel you have achieved sufficient mastery to answer multiple-choice questions correctly?

- For which content topics discussed in this chapter do you feel you have achieved sufficient mastery to discuss effectively in a free response?

- For which content topics discussed in this chapter do you feel you need more work before you can answer multiple-choice questions correctly?

- For which content topics discussed in this chapter do you feel you need more work before you can discuss effectively in a free response?

- What parts of this chapter are you going to re-review?

- Will you seek further help, outside this book (such as a teacher, tutor, or AP Central), on any of the content in this chapter—and if so, on what content?

# Chapter 7
# Photosynthesis

Plants and algae are producers. All they do is bask in the sun, churning out the glucose necessary for life.

As we discussed earlier, photosynthesis is the process by which light energy is converted to chemical energy. Here's an overview of photosynthesis:

$$6CO_2 + 6H_2O \xrightarrow{\text{sunlight}} C_6H_{12}O_6 + 6O_2$$

You'll notice from this equation that carbon dioxide and water are the raw materials plants use to manufacture simple sugars. But remember, there's much more to photosynthesis than what's shown in the simple reaction above. We'll soon see that this beautifully orchestrated process occurs thanks to a whole host of special enzymes and pigments. But before we turn to the stages in photosynthesis, let's talk about where photosynthesis occurs.

## THE ANATOMY OF A LEAF

Photosynthesis occurs in the leaves of plants. Here's a cross-sectional view of a typical leaf:

If you look closely at the leaf of any plant, the first thing you'll notice is the waxy covering called the cuticle. The **cuticle** is produced by the **upper epidermis** to protect the leaf from water loss through evaporation. Just below the upper epidermis is the **palisade parenchyma**. These cells contain lots of chloroplasts, which are the primary sites of photosynthesis.

Now let's look at an individual chloroplast. If you split the membrane of a chloroplast, you'll find a fluid-filled region called the **stroma**. Inside the stroma are structures that look like stacks of coins. These structures are the **grana**.

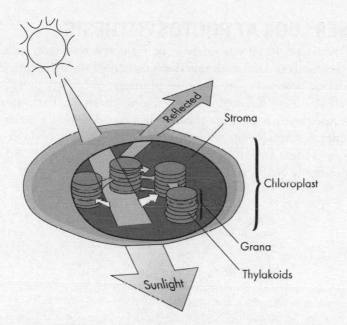

The many disk-like structures that make up the grana are called the **thylakoids.** They contain chlorophyll, the light-absorbing pigment that drives photosynthesis, as well as enzymes involved in the process.

Now we'll discuss some structures that are not involved in photosynthesis but are important for the AP test. Just below the palisade parenchyma, you'll find irregular-shaped cells called **spongy parenchyma**. They allow for diffusion of gases, especially $CO_2$, within the leaf. The **vascular bundles** are found in this layer of the leaf. The vascular bundles include **xylem** and **phloem**, "vessels" that transport materials throughout the plant. At the **lower epidermis** are tiny openings called **stomates,** which allow for gas exchange and transpiration. Surrounding each stomate are **guard cells**, which control the opening and closing of the stomates.

## A CLOSER LOOK AT PHOTOSYNTHESIS

There are two stages in photosynthesis: the **light reaction** (also called the light-dependent reaction) and the **dark reaction** (also called the light-independent reaction). The whole process begins when the **photons** (or "energy units") of sunlight strike a leaf, activating chlorophyll and exciting electrons. The activated chlorophyll molecule then passes these excited electrons down to a series of electron carriers, ultimately producing ATP and NADPH. The whole point of the light reaction is to produce two things: (1) energy in the form of ATP and (2) electron carriers, specifically NADPH.

Both of these products, along with carbon dioxide, are then used in the dark reaction (light-independent) to make carbohydrates.

# The Light Reaction

Many light-absorbing pigments participate in photosynthesis. Some of the more important ones are **chlorophyll *a***, **chlorophyll *b***, and **carotenoids**. These pigments are clustered in the thylakoid membrane into units called antenna complexes.

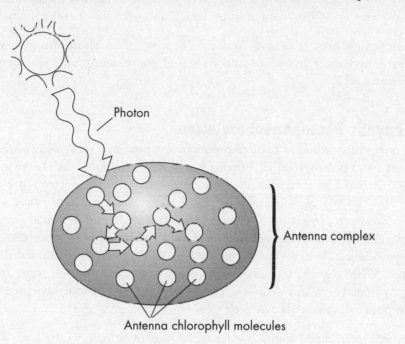

All of the pigments within a unit are able to "gather" light, but they're not able to "excite" the electrons. Only one special molecule—located in the **reaction center**—is capable of transforming light energy to chemical energy. In other words, the other pigments, called **antenna pigments**, "gather" light and "bounce" the energy to the reaction center.

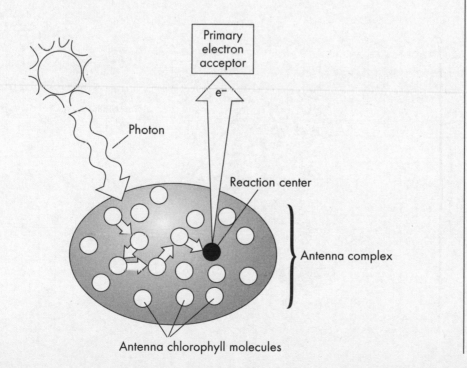

There are two types of reaction centers: **photosystem I (PS I)** and **photosystem II (PS II)**. The principal difference between the two is that each reaction center has a specific type of chlorophyll—chlorophyll *a*—that absorbs a particular wavelength of light. For example, **P680**, the reaction center of photosystem II, has a maximum absorption at a wavelength of 680 nanometers. The reaction center for photosystem I—**P700**—best absorbs light at a wavelength of 700 nanometers.

When light energy is used to make ATP, it is called **photophosphorylation**. We're using light (that's *photo*) and ADP and phosphates (that's *phosphorylation*) to produce ATP.

## Noncyclic Photophosphorylation

The noncyclic method of photophosphorylation produces ATP using both photosystem I and photosystem II. When a leaf captures sunlight, the energy is sent to P680, the reaction center for photosystem II. The activated electrons are trapped by P680 and passed to a molecule called the primary acceptor. They are then passed down to carriers in the electron transport chain and eventually enter photosystem I. Some of the energy that dissipates as electrons move along the chain of acceptors will be used to "pump" protons across the membrane into the thylakoid lumen. When the reaction center P680 absorbs light, it also splits water into oxygen, hydrogen ions, and electrons. That process is called **photolysis**. The electrons from photolysis replace the missing electrons in photosystem II.

Remember the chemiosmotic theory mentioned in aerobic respiration? Well, the same mechanism applies to photosynthesis. Here's how it works: Hydrogen ions accumulate inside the thylakoids when photolysis occurs. A proton gradient is established. As the hydrogen ions move through ATP synthase, ADP and $P_i$ produce ATP.

When these electrons in photosystem I receive a second boost, they're activated again. The electrons are passed through a second electron transport chain until they reach the final electron acceptor NADP⁺ to make **NADPH**.

## Cyclic Photophosphorylation

The cyclic method uses a much simpler pathway to generate ATP. The electrons in photosystem I are excited and leave the reaction center, P700. They are passed from carrier to carrier in the electron transport system and eventually return to P700:

At the end of this cycle, only ATP is produced. This pathway is called cyclic photophosphorylation because the electrons from P700 return to the same reaction center. Unfortunately, this method isn't as efficient as the noncyclic pathway since it doesn't produce NADPH. Plants use this method only when there aren't enough NADP molecules to accept electrons. Keep in mind:

> The light reaction occurs in the thylakoids.

Let's review the cyclic and noncyclic phosphorylation steps of the light reaction:

**Noncyclic phosphorylation**

- P680 in photosystem II captures light and passes excited electrons down an electron transport chain to produce ATP.

- P700 in photosystem I captures light and passes excited electrons down an electron transport chain to produce NADPH.

- A molecule of water is split by sunlight, releasing electrons, hydrogen, and free $O_2$.

**Cyclic phosphorylation**

- P700 in photosystem I captures light and passes excited electrons down an electron transport chain to produce ATP.

- NADPH is not produced.

- Water is not split by sunlight.

Both reactions occur in the grana of chloroplasts, where the thylakoids are found. Remember: The light-absorbing pigments and enzymes for the light-dependent reactions are found within the thylakoids.

# The Light-Independent Reaction

Now let's turn to the dark reaction. The dark reaction uses the products of the light reaction—ATP and NADPH—to make sugar. We now have energy to make glucose, but what do plants use as their carbon source? $CO_2$, of course. You've probably heard of the term **carbon fixation**. All this means is that $CO_2$ from the air is converted into carbohydrates. This step occurs in the stroma of the leaf.

## The Calvin Cycle: The $C_3$ Pathway

We're finally ready to make glucose. $CO_2$ enters the **Calvin cycle** and combines with a 5-carbon molecule called **ribulose bisphosphate (RuBP)** to make an unstable six-carbon compound. The enzyme RuBP carboxylase, or *rubisco*, catalyzes this reaction.

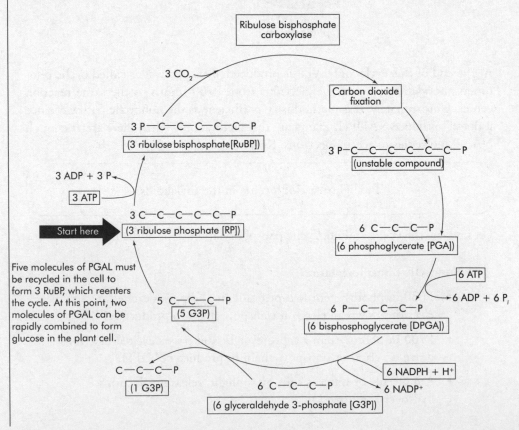

The easiest way to view the Calvin cycle is to consider 6 RuBP and 6 $CO_2$ at the start. Next, 12 ATP and 12 NADPH are used to convert 12 PGA to 12 G3P, an energy-rich molecule. ADP and NADP$^+$ are released and then recycled into the thylakoid where they will again be available for the light-dependent reactions. Two of the G3P are used to make glucose while the remaining 10 are rearranged into 6 RuBPs ready for the next round of the cycle. Since G3P, a three-carbon molecule, is the first stable product, this method of producing glucose is called the **C$_3$ pathway**.

The following table summarizes the stages of photosynthesis:

| Stage of Photosynthesis | Location | Input | Output |
|---|---|---|---|
| **Light Dependent Reactions** Noncyclic Flow | Thylakoid membrane of chloroplasts Photosystem II (P680) Photosystem I (P700) | Photons $H_2O$ | NADPH ATP $O_2$ |
| **Light Dependent Reactions** Cyclic Flow | Thylakoid membrane of chloroplasts Photosystem I (P700) | Photons | ATP |
| **Light Independent Reactions** Calvin Cycle | Stroma | 3 $CO_2$ 3 RuBP 9 ATP 6 NADPH | 1 glyceraldehyde-3-phosphates (G3P) |

## Photorespiration

Sometimes intensely bright light tends to stunt the growth of C$_3$ plants. Why? Lighted conditions can trigger a process called **photorespiration**. Photorespiration is the pathway that leads to the fixation of oxygen. During this process, RuBP carboxylase reduces the $CO_2$ concentration to the point that it starts incorporating $O_2$ instead. This process makes $CO_2$-fixing less efficient.

## C$_4$: The Alternative Pathway

The C$_3$ pathway is not the only way to "fix" $CO_2$. Some plants, such as sugar cane and corn, have a more efficient way to fix carbon dioxide. In these plants, carbon dioxide first "combines" with **phosphoenolpyruvate (PEP)** in mesophyll cells to form oxaloacetate, a four-carbon molecule. The enzyme that fixes PEP, **PEP carboxylase**, has a high affinity for $CO_2$, even under unusually low concentrations. Oxaloacetate is then converted to malate. Malate enters the bundle sheath cells, a tissue surrounding the leaf vein, and is converted to pyruvate and carbon dioxide. Carbon dioxide is then released for uptake into the regular Calvin cycle to make glucose.

The **C$_4$ pathway** works particularly well for plants found in hot, dry climates. It enables them to fix $CO_2$, even when the supply is greatly diminished. For our purposes, just remember that the subscripts in both C$_3$ and C$_4$ refer to the number of carbons *initially* involved in making sugar. However, both pathways ultimately use the Calvin cycle to produce glucose.

Let's summarize the important facts about the dark reaction:

- The Calvin cycle occurs in the stroma of chloroplasts.
- ATP and NADPH from the light reaction are necessary for carbon fixation.
- $CO_2$ is fixed to form glucose.

## CAM Photosynthesis

In some plants, the stomates are closed during the day to reduce excessive water loss from transpiration. You might think that this would prevent these plants from carrying out photosynthesis during the day. However, desert plants have evolved a way to perform photosynthesis when their stomates are closed; it's called **CAM (crassulacean acid metabolism) photosynthesis**. In CAM photosynthesis, a process similar to C$_4$ photosynthesis, PEP carboxylase is used to fix $CO_2$ to oxaloacetate, but oxaloacetate is converted to malic acid instead of malate (an ionized form of malate) and sent to the cell's vacuole. During the day, malic acid is converted back to oxaloacetate and carbon dioxide is released for photosynthesis. $CO_2$ then enters the Calvin cycle.

## KEY TERMS

cuticle

upper epidermis

palisade parenchyma

stroma

grana

thylakoids

spongy parenchyma

vascular bundles

xylem

phloem

lower epidermis

stomates

guard cells

light reaction (light-dependent reaction)

dark reaction (light-independent reaction)

photons

chlorophyll *a*

chlorophyll *b*

carotenoids

reaction center

antenna pigments

photosystem I (PS I)

photosystem II (PS II)

P680

P700

photophosphorylation

photolysis

NADPH

noncyclic photophosphorylation

cyclic photophosphorylation

carbon fixation

Calvin cycle

ribulose bisphosphate (RuBP)

$C_3$ pathway

photorespiration

phosphoenolpyruvate (PEP)

PEP carboxylase

$C_4$ pathway

CAM (crassulacean acid metabolism) photosynthesis

# Summary

- Plants are the producers of an ecosystem, and they harness light energy from the sun and $CO_2$ from the air to produce chemical energy in their chloroplasts.

- Light-dependent reactions use the photosystems II and I to harvest photons and produce high-energy electrons that are carried by NADPH into the stroma of the chloroplast. Passage of high-energy electrons through an electron transport chain in the thylakoid membrane generates a proton gradient that drives ATP synthesis on the stroma side of the membrane.

- In the stroma, the Calvin cycle fixes $CO_2$ from the air with ribulose bisphosphate carboxylase (Rubisco) enzyme. Through many steps in this cycle and the use of many NADPH and ATP molecules, $CO_2$ is used to store chemical energy and bring new carbons into the cell that can be used for subsequent metabolic processes.

- Other techniques for carbon fixation, such as the $C_4$ pathway and CAM photosynthesis, use alternative methods, which help plants living in extreme habitats to optimize photosynthesis without losing water.

# Chapter 7 Drill

Answers and explanations can be found in Chapter 16.

1. Antenna complexes are located on the surface of thylakoid membranes and play an important role in photosynthesis. Surrounding the reaction center of an antenna complex are many antenna pigments. What is the functional role of these antenna pigments?

   (A) They are photosystems, which harness light and directly generate ATP by photophosphorylation.
   (B) They absorb and harness green light to ensure that only the correct wavelengths of light reach the reaction center.
   (C) They absorb light and transfer its energy to the reaction center.
   (D) They produce chlorophyll *a* and chlorophyll *b* to be used by the reaction centers for photophosphorylation.

2. The light-dependent and light-independent stages of photosynthesis are both linked together. Which of the following is generated by the light-independent reactions and is used in the light-dependent reactions?

   (A) ATP
   (B) $NAD^+$
   (C) NADPH
   (D) $NADP^+$

3. Carbon dioxide generated as a by-product of cellular respiration is recycled during photosynthesis. During which process of photosynthesis is carbon dioxide consumed?

   (A) Cyclic photophosphorylation
   (B) Noncyclic photophosphorylation
   (C) The $C_3$ pathway
   (D) Krebs cycle

Noncyclic photophosphorylation (shown above) is a critical process by which photophosphorylation of ATP occurs through the use of both photosystems I and II. During noncyclic photophosphorylation, light activates photosystem II that donates electrons to the primary acceptor molecule, beginning an electron transport chain cascade that ultimately generates ATP and activates photosystem I. Photosystem I donates electrons to another primary acceptor that can be used to generate NADPH. Although this pathway is more complex than cyclic photophosphorylation, it is the preferred mechanism of photophosphorylation in the chloroplast.

4.   Why is noncyclic photophosphorylation preferred over cyclic photophosphorylation, despite being more complex?

(A) Cyclic photophosphorylation fails to produce ATP, instead producing lower energy ADP molecules.
(B) Cyclic photophosphorylation fails to produce NADPH, which is necessary for the light-independent pathway.
(C) Cyclic photophosphorylation fails to recycle $CO_2$ gas generated as a byproduct of cellular respiration in the mitochondria.
(D) Cyclic photophosphorylation requires more energy input than noncyclic photophosphorylation.

5.   Suppose a scientist treats a plant with a ferredoxin inhibitor. Which of the following molecules may still be generated under these conditions?

(A) Glucose
(B) NADPH
(C) RuBP
(D) $O_2$

6.   Which of the following describes a key difference between cyclic photophosphorylation and noncyclic phosphorylation?

(A) Noncyclic photophosphorylation splits water by sunlight, whereas cyclic photophosphorylation does not.
(B) Cyclic photophosphorylation uses P680 in photosystem II to produce ATP, whereas noncyclic photophosphorylation uses P700 in photosystem I.
(C) Cyclic photophosphorylation uses P700 in photosystem I to produce ATP, whereas noncyclic photophosphorylation uses P680 in photosystem II.
(D) Cyclic photophosphorylation generates NADPH using photosystem I, whereas no NADPH is generated during noncyclic photophosphorylation.

7. Which of the following is accurate, considering the ATP produced during the light reactions of photosynthesis?

   (A) ATP is generated in the thylakoid lumen by ATP-synthase.
   (B) ATP generated by the light reactions is used primarily in the Calvin cycle.
   (C) ATP is necessary to generate NADPH.
   (D) ATP cannot be generated in chloroplasts without oxygen present.

8. Which is an accurate comparison of the electrochemical gradients established by the electron transport chain in chloroplasts and in mitochondria?

   (A) In mitochondria, protons are pumped across the inner mitochondrial membrane in order to generate ATP, but in chloroplasts, protons are pumped across the thylakoid membrane to generate NADPH.
   (B) In mitochondria, ATP synthesis occurs in the matrix, but in chloroplasts, ATP synthesis occurs in the lumen.
   (C) In chloroplasts, protons are pumped against their concentration gradient, but in mitochondria, protons are pumped down their concentration gradient.
   (D) In mitochondria, protons are pumped into the inner mitochondrial membrane space and ATP is generated in the matrix, but in chloroplasts, protons are pumped into the thylakoid membrane space and ATP is generated in the stroma.

9. Algae cells growing in normal sunlight conditions were moved to an airtight container and treated with only green light for one week. The oxygen content of the water was measured at the end of the treatment period. Predict the difference between the control sample grown in sunlight and the sample grown under green light.

   (A) The oxygen concentration was higher in the sample grown under green light.
   (B) The oxygen concentration was higher in the sample grown under sunlight.
   (C) The oxygen concentration was the same in both samples.
   (D) The carbon dioxide concentration was higher in the sample grown under sunlight.

10. Photorespiration results when oxygen is bound by rubisco instead of carbon dioxide. This process depletes ATP in the plant because the Calvin cycle continues without adding any carbon to the cells. Several different plants were placed together in a chamber with high temperature, high intensity light, water, oxygen, and carbon dioxide. The rates of photorespiration were measured. Which plant would expect to have the highest levels of photorespiration?

   (A) Cactus, a $C_4$ plant
   (B) Pineapple plant, a CAM plant
   (C) Daisies, a $C_3$ plant
   (D) Kelp, a plant with $C_3$ and $C_4$ plant capabilities

## REFLECT

Respond to the following questions:

- For what content topics discussed in this chapter do you feel you have achieved sufficient mastery to answer multiple-choice questions correctly?

- For which content topics discussed in this chapter do you feel you have achieved sufficient mastery to discuss effectively in a free response?

- For which content topics discussed in this chapter do you feel you need more work before you can answer multiple-choice questions correctly?

- For which content topics discussed in this chapter do you feel you need more work before you can discuss effectively in a free response?

- What parts of this chapter are you going to re-review?

- Will you seek further help, outside this book (such as a teacher, tutor, or AP Central), on any of the content in this chapter—and if so, on what content?

# Chapter 8
# Molecular Genetics

## DNA: THE BLUEPRINT OF LIFE

All living things possess an astonishing degree of organization. From the simplest single-celled organism to the largest mammal, millions of reactions and events must be coordinated precisely for life to exist. This coordination is directed from the nucleus of the cell, by deoxyribonucleic acid, or DNA. DNA is the hereditary blueprint of the cell.

The DNA of a cell is contained in structures called chromosomes. Chromosomes consist of DNA wrapped around proteins called histones. When the genetic material is in a loose form in the nucleus it is called **euchromatin,** and its genes are active, or available for transcription. When the genetic material is fully condensed into coils it is called **heterochromatin,** and its genes are generally inactive. Situated in the nucleus, chromosomes direct and control all the processes necessary for life, including passing themselves and their information on to future generations. In this chapter, we'll look at precisely how they accomplish this.

## THE MOLECULAR STRUCTURE OF DNA

The DNA molecule consists of two strands that wrap around each other to form a long, twisted ladder called a **double helix**. The structure of DNA was brilliantly deduced in 1956 by two scientists named Watson and Crick.

DNA is made up of repeated subunits of **nucleotides**. Each nucleotide has a **five-carbon sugar,** a **phosphate**, and a **nitrogenous base**. Take a look at the nucleotide below. This particular nucleotide contains a nitrogenous base called **adenine**:

The name of the pentagon-shaped sugar in DNA is **deoxyribose**. Hence, the name *deoxyribo*nucleic acid. Notice that the sugar is linked to two things: a phosphate and a nitrogenous base. A nucleotide can have one of four different nitrogenous bases:

- adenine—a purine (double-ringed nitrogenous base)

- guanine—a purine (double-ringed nitrogenous base)

- cytosine—a pyrimidine (single-ringed nitrogenous base)

- thymine—a pyrimidine (single-ringed nitrogenous base)

Any of these four bases can attach to the sugar. As we'll soon see, this is extremely important when it comes to the "sense" of the genetic code in DNA.

The nucleotides can link up in a long chain to form a single strand of DNA. Here's a small section of a DNA strand:

The nucleotides themselves are linked together by **phosphodiester bonds**.

## Two DNA Strands

Now let's look at the way two DNA strands get together. Again, think of DNA as a ladder. The sides of the ladder consist of alternating sugar and phosphate groups, while the rungs of the ladder consist of a pair of nitrogenous bases.

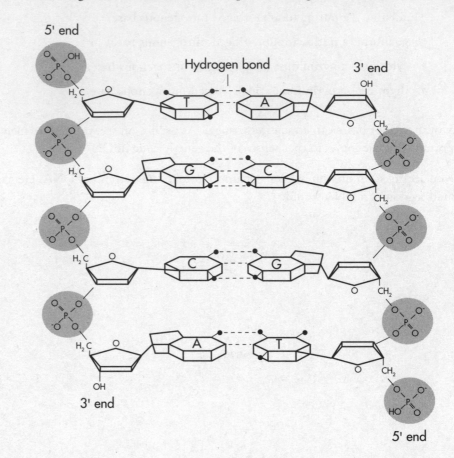

The nitrogenous bases pair up in a particular way. Adenine in one strand always binds to thymine (A–T or T–A) in the other strand. Similarly, guanine always binds to cytosine (G–C or C–G). This predictable matching of the bases is known as **base pairing**.

The two strands are said to be **complementary**. This means that if you know the sequence of bases in one strand, you'll know the sequence of bases in the other strand. For example, if the base sequence in one DNA strand is A–T–C, the base sequence in the complementary strand will be T–A–G.

The two DNA strands run in opposite directions. You'll notice from the figure above that each DNA strand has a 5' end and a 3' end, so-called for the carbon that ends the strand (i.e., the fifth carbon in the sugar ring is at the 5' end, while the third carbon is at the 3' end). The 5' end has a phosphate group and the 3' end has an OH, or "hydroxyl," group. The 5' end of one strand is always opposite to the 3' end of the other strand. The strands are therefore said to be **antiparallel**.

The DNA strands are linked by **hydrogen bonds**. Two hydrogen bonds hold adenine and thymine together and three hydrogen bonds hold cytosine and guanine together.

Before we go any further, let's review the base pairing in DNA.

- Adenine pairs up with thymine (A–T or T–A) by forming two hydrogen bonds.
- Cytosine pairs up with guanine (G–C or C–G) by forming three hydrogen bonds.

## WHY DNA IS IMPORTANT

You'll recall from our discussion of bioenergetics in Chapter 6 that enzymes are proteins that are essential for life. This is true not only because they help liberate energy stored in chemical bonds, but also because they direct the construction of the cell. This is where DNA comes into the picture.

DNA's main role is directing the manufacture of proteins. These proteins, in turn, regulate everything that occurs in the cell. But DNA does not *directly* manufacture proteins. Instead, DNA passes its information to an intermediate molecule known as ribonucleic acid (RNA). These RNA molecules carry out the instructions in DNA, producing the proteins that determine the course of life.

The flow of genetic information is therefore:

$$DNA \rightarrow RNA \rightarrow proteins$$

This is the central dogma of molecular biology.

We said in the beginning of this chapter that DNA is the hereditary blueprint of the cell. By directing the manufacture of proteins, DNA serves as the cell's blueprint. But how is DNA inherited? For the information in DNA to be passed on, it must first be copied. This copying of DNA is known as **DNA replication**.

## DNA REPLICATION

Because the DNA molecule is twisted over on itself, the first step in replication is to unwind the double helix by breaking the hydrogen bonds. This is accomplished by an enzyme called **helicase**. The exposed DNA strands now form a *y*-shaped **replication fork**:

Now each strand can serve as a template for the synthesis of another strand. DNA replication begins at specific sites called **origins of replication**. Because the DNA helix twists and rotates during DNA replication, another class of enzymes, called DNA **topoisomerases**, cuts and rejoins the helix to prevent tangling. The enzyme that performs the actual addition of nucleotides alongside the naked strand is **DNA polymerase**. But DNA polymerase, oddly enough, can only add nucleotides to the 3' end of an existing strand. Therefore, to start off replication at the 5' end, DNA polymerase must add nucleotides to an **RNA primase**—a short strand of RNA nucleotides. The primer is later degraded by enzymes, and the space is filled with DNA.

One strand is called the **leading strand**, and it is made continuously. That is, the nucleotides are steadily added one after the other by DNA polymerase. The other strand—the **lagging strand**—is made discontinuously. Unlike the leading strand, the lagging strand is made in pieces of nucleotides known as **Okazaki fragments**. Why is the lagging strand made in small pieces?

Normally, nucleotides are added only in the 5' to 3' direction. However, when the double-helix is "unzipped," one of the two strands is oriented in the opposite direction—3' to 5'. Because DNA polymerase doesn't work in this direction, nucleotides—the Okazaki fragments—need to be added in pieces. These fragments are eventually linked together by the enzyme **DNA ligase** to produce a continuous strand. Finally, hydrogen bonds form between the new base pairs, leaving two identical copies of the original DNA molecule.

When DNA is replicated, we don't end up with two entirely new molecules. Each new molecule has half of the original molecule. Because DNA replicates in this way, by conserving half of the original molecule in each of the two new ones, it is said to be **semiconservative**.

Many enzymes and proteins are involved in DNA replication. The ones you'll need to know for the AP Biology Exam are DNA helicase, DNA polymerase, DNA ligase, topoisomerase, and RNA primase.

- **Helicase** unwinds our double helix into two strands.
- **Polymerase** adds nucleotides to an existing strand.
- **Ligase** brings together the Okazaki fragments.
- **Topoisomerase** cuts and rejoins the helix.
- **RNA primase** catalyzes the synthesis of RNA primers.

**Test Tip**
Of the enzymes and proteins involved in DNA replication, you're expected to know five: helicase, polymerase, ligase, topoisomerase, and RNA primase.

# RNA

Now that we've seen how DNA is replicated, let's take a look at how the genetic code is expressed as proteins. As we mentioned earlier, genetic information is first passed to an intermediate molecule called RNA. Proteins called **transcription factors** control the transfer of genetic information from DNA to RNA by binding to specific DNA sequences. Here's a "roadmap" of how information is transferred from DNA to proteins:

$$\text{DNA} \xrightarrow[\text{in the nucleus}]{\text{transcription}} \text{RNA} \xrightarrow[\text{in the cytoplasm}]{\text{translation}} \text{Proteins}$$

Before we discuss what RNA does, let's talk about its structure. Although RNA is also made up of nucleotides, it differs from DNA in three ways:

1. RNA is single-stranded, not double-stranded.
2. The five-carbon sugar in RNA is **ribose** instead of deoxyribose.
3. The RNA nitrogenous bases are adenine, guanine, cytosine, and a different base called **uracil**. Uracil replaces thymine.

Here's a table to compare DNA and RNA. Keep these differences in mind—ETS loves to test you on them.

| Differences Between DNA and RNA | | |
|---|---|---|
| | **DNA (double-stranded)** | **RNA (single-stranded)** |
| Sugar: | deoxyribose | ribose |
| Bases: | adenine | adenine |
| | guanine | guanine |
| | cytosine | cytosine |
| | thymine | uracil |

There are three types of RNA: messenger RNA (mRNA), ribosomal RNA (rRNA), and transfer RNA (tRNA). All three types of RNA are key players in the synthesis of proteins.

- **Messenger RNA (mRNA)** copies the information stored in the strand of DNA and carries it to the ribosome.

- **Ribosomal RNA (rRNA)**, which is produced in the nucleolus, makes up part of the ribosomes. You'll recall from our discussion of the cell in Chapter 5 that the ribosomes are the sites of protein synthesis. We'll see how they function a little later on.

- **Transfer RNA (tRNA)** shuttles amino acids to the ribosomes. It is responsible for bringing the appropriate amino acids into place at the appropriate time. It does this by reading the message carried by the mRNA.

Now that we know about the different types of RNA, let's see how they direct the synthesis of proteins.

## PROTEIN SYNTHESIS

**Protein synthesis** involves three basic steps: **transcription**, **RNA processing**, and **translation**.

## Transcription

Transcription involves copying the genetic code directly from DNA. The initial steps in transcription are similar to the initial steps in DNA replication. The obvious difference is that, whereas in replication we end up with a complete copy of the cell's DNA, in transcription we end up with only a partial copy in the form of mRNA.

Transcription involves three phases: initiation, elongation, and termination. As in DNA replication, the first initiation step in transcription is to unwind and unzip the DNA strands using helicase. Transcription begins at special sequences of the DNA strand called **promoters**. Because RNA is single-stranded, we have to copy only *one* of the two DNA strands. The strand that serves as the template is the **sense strand**. The other strand that lies dormant is the **antisense strand**.

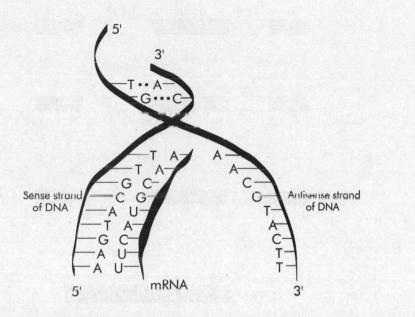

This time, RNA nucleotides line up alongside one DNA strand to form an mRNA strand. Another enzyme, **RNA polymerase**, brings free-floating RNA nucleotides to the DNA strand. As we saw earlier, guanines and cytosines pair up. However, the RNA molecule replaces the DNA's thymine molecules with uracils. In other words, the exposed adenines are now paired up with uracil instead of thymine.

Once the mRNA finishes adding on nucleotides and reaches the termination sequence, it separates from the DNA template, completing the process of transcription. The new RNA has now transcribed, or "copied," the sequence of nucleotide bases directly from the exposed DNA strand.

## RNA Processing

Now the mRNA strand is ready to move out of the nucleus. But before the mRNA molecule can leave the nucleus, it must be processed. That means the mRNA has to be modified before it exits the nucleus.

A newly made mRNA molecule (called the heterogeneous nuclear RNA, or hnRNA) contains more nucleotides than it needs to code for a protein. The mRNA consists of both coding regions and noncoding regions. The regions that express the code for the polypeptide are **exons**. The noncoding regions in the mRNA are **introns**.

The introns—the intervening sequences—must be removed before the mRNA leaves the nucleus. The removal of introns is a complex process accomplished by an RNA-protein complex called a **spliceosome**. In addition, a **poly(A) tail** is added to the 3' end and a **5' GTP cap** is added to the 5' end. This process produces a final mRNA that is shorter than the transcribed mRNA.

## Translation

Now, the mRNA leaves the nucleus and searches for a ribosome. The mRNA molecule carries the message from DNA in the form of **codons**, a group of three bases, or "letters," that corresponds to one of 20 amino acids. The genetic code is redundant, meaning that certain amino acids are specified by more than one codon.

The mRNA attaches to the ribosome to initiate translation and "waits" for the appropriate amino acids to come to the ribosome. That's where tRNA comes in. A tRNA molecule has a unique three-dimensional structure that resembles a four-leaf clover:

One end of the tRNA carries an amino acid. The other end, called an **anticodon**, has three nitrogenous bases that can base pair with the codon in the mRNA.

Transfer RNAs are the "go-betweens" in protein synthesis. Each tRNA becomes charged and enzymatically attaches to an amino acid in the cell's cytoplasm and "shuttles" it to the ribosome. The charging enzymes involved in forming the bond between the amino acid and tRNA require ATP.

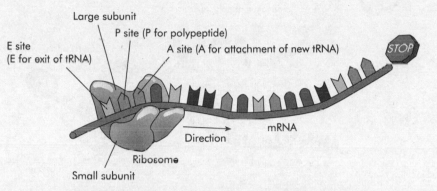

Note: Each site E, P, or A are grooves for the tRNA to fit in.

Translation also involves three phases: initiation, elongation, and termination. Initiation begins when a ribosome attaches to the mRNA.

What does the ribosome do? It helps the process along by holding everything in place while the RNAs assist in assembling polypeptides.

## Initiation

Ribosomes contain three binding sites: an **A site**, a **P site**, and an **E site**. An initiator tRNA serves to activate translation and occupies the P site. In all organisms the codon for the initiation of protein synthesis is A–U–G, which codes for the amino acid methionine. The tRNA with the complementary anticodon, U–A–C, is methionine's personal shuttle. Once the methionine tRNA is attached to the P site, the A site can be filled by the appropriate tRNA that corresponds to the next codon. The E site binds a free tRNA before it exits the ribosome.

## Elongation

The addition of amino acids is called elongation. Remember that the mRNA contains many codons, or "triplets," of nucleotide bases. As each amino acid is brought to the mRNA, it is linked to its neighboring amino acid by a peptide bond. When many amino acids link up, a polypeptide is formed.

## Termination

How does this process know when to stop? The synthesis of a polypeptide is ended by **stop codons**. A codon doesn't always code for an amino acid; there are three that serve as a stop codon. Termination occurs when the ribosome runs into one of these three stop codons.

# Higher Protein Structure

The polypeptide has to go through several changes before it can officially be called a protein. Proteins can have four levels of structure. The linear sequence of the amino acids is called the **primary structure** of a protein. Now the polypeptide begins to twist, forming either a coil (known as an alpha helix) or zigzagging pattern (known as beta-pleated sheets). These are examples of proteins' **secondary structures**. Next, the polypeptide folds in a three dimensional pattern. This is called the **tertiary structure**. Finally, when two or more polypeptides get together, we call it a **quaternary structure**. Once that's complete the protein is ready to perform its task.

One more thing: In some cases, the folding of proteins involves other proteins known as **chaperone proteins** (or **chaperonins**). They help the protein fold properly and make the process more efficient.

How about a little review?

- In transcription, mRNA copies, or "transcribes," the code from an exposed strand of DNA in the nucleus.

- The mRNA is "processed" by having its introns, or noncoding sequences, removed.

- Now, ready to be translated, mRNA proceeds to the ribosome.

- Free-floating amino acids are picked up by tRNA and shuttled over to the ribosome, where mRNA awaits.

- In translation, the anticodon of a tRNA molecule carrying the appropriate amino acid base pairs with the codon on the mRNA.

- As new tRNA molecules match up to new codons, the ribosome holds them in place, allowing peptide bonds to form between the amino acids.
- The newly formed polypeptide grows until a stop codon is reached.
- The polypeptide or protein is released into the cell.

# MUTATIONS

There are two types of mutations: base substitutions and gene rearrangements.

## Base Substitution

Base substitution (point) mutations result when one base is substituted for another. Point mutations within a gene can be **nonsense mutations** (early termination of protein synthesis), **missense mutations** (a codon is altered and produces a different amino acid), or **silent mutations** (a nucleotide is selected which creates a codon that codes for the same amino acid and therefore does not change the corresponding protein sequence).

## Gene Rearrangements

Gene rearrangements involve DNA sequences that have deletions, duplications, inversions, and translocations which are often caused by chemical and radioactive agents.

1. Insertions and **deletions** result in the gain or loss, respectively, of DNA or a gene. They can either involve the addition (insertion) or removal (deletion) of as little as a single base or much larger sequences of DNA. Insertions and deletions may have devastating consequences during protein translation because the introduction or deletion of bases often results in a change in the sequence of codons used by the ribosome (called a frameshift) to synthesize a polyprotein.

   Example:

   | Original DNA Sequence: | ATG | TAT | AAA | CAT | TGA |
   |---|---|---|---|---|---|
   | Original Polyprotein: | Met | Tyr | Lys | His | Stop |
   | DNA Sequence after Insertion: | ATG | CTA | TAA | ACA | TTG A |
   | Polyprotein Sequence after Insertion: | Met | Leu | Stop | - | - |

   In this example, the insertion of an additional cytosine nucleotide resulted in a frameshift and a premature stop codon.

2. **Duplications** can result in an extra copy of genes and are usually caused by unequal crossing over during meiosis or chromosome rearrangements. This may result in new traits because one copy can maintain the gene's original function and one copy may evolve a new function.

3. **Inversions** can result when changes occur in the orientation of chromosomal regions. This may cause harmful effects if the inversion involves a gene or an important sequence involved in the regulation of **gene expression.**

4. **Translocations** occur when a portion of two different chromosomes (or a single chromosome in two different places) breaks and rejoins in a way that the DNA sequence or gene is lost, repeated, or interrupted.

Either type of mutation, if not repaired, will be kept in subsequent rounds of replication. Mutations in somatic cells may damage the cell, make it cancerous, or kill it. Mutations in a germ cell will be passed down to the next generation.

# RECOMBINANT DNA

**Recombinant DNA** is generated by combining DNA from multiple sources to create a unique DNA that is not found in nature. A common application of recombinant DNA technology is the introduction of a eukaryotic gene of interest (such as insulin) into a bacterium for production. This branch of technology that produces new organisms or products by transferring genes between cells is called **genetic engineering**.

How does this process work? Unique sequences, which flank the gene of interest in eukaryotic DNA, are recognized and cut by special enzymes called restriction enzymes. Using the same **restriction enzymes**, a cut is made in a circular non-essential bacterial DNA called a **plasmid** to create a site for insertion of the eukaryotic DNA. Once the plasmid has been cut, the eukaryotic DNA fragment containing the gene of interest is inserted into the plasmid by the enzyme DNA ligase.

In order to introduce the recombinant DNA plasmid containing the eukaryotic gene of interest into a bacterial cell, the plasmid DNA is combined with bacteria and placed under conditions that favor uptake of the DNA. This process is called **transformation**. Once transformed, these bacteria may be expanded, and the gene, or protein of interest, is purified in large quantity.

The DNA fragments can be separated according to their molecular weight using **gel electrophoresis**. Because DNA and RNA are negatively charged, they migrate through the gel toward the positive pole of the electrical field. The smaller the fragments, the faster they move through the gel. Restriction enzymes are also used to create a molecular fingerprint. When restriction fragments between individuals of the same species are compared, the fragments differ in length because of polymorphisms, which are slight differences in DNA sequences. These fragments are called **restriction fragment length polymorphisms**, or RFLPs. In **DNA fingerprinting**, RFLPs produced from DNA left at a crime scene are compared to RFLPs from the DNA of suspects.

# BIOTECHNOLOGY

## Polymerase Chain Reaction (PCR)

A few years ago, it would take weeks of tedious experiments to identify and study specific genes. Today, thanks to **polymerase chain reaction (PCR)**, we are able to make billions of identical copies of genes within a few hours. To do PCR, the process of DNA replication is slightly modified. In a small PCR tube, DNA, primers, *Taq* Polymerase, and lots of DNA nucleotides (A's, C's, G's, and T's) are mixed together.

In a PCR machine, or thermocycler, the tube is heated, cooled, and warmed many times. Each time the machine is heated, the hydrogen bonds break, separating the double-stranded DNA. As it is cooled, the primers bind to the sequence flanking the region of the DNA we want to copy. When it is warmed, *Taq* Polymerase binds to the primers on both strands and adds nucleotides on each template strand. After this first cycle is finished, there are two identical double-stranded DNA molecules. When the second cycle is completed, these two double-stranded DNA segments will have been copied into four. The process repeats itself over and over, creating as much DNA as needed.

Today, a thermocycler is commonplace in science labs. It is regularly used to study small amounts of DNA from crime scenes, determine the origin of our foods, detect diseases in animals and humans, and to better understand the inner workings of our cells.

## Transformation

Insulin, the protein hormone that lowers blood sugar levels, can now be made for medical purposes by bacteria. Yes, bacteria can be induced to use the universal DNA code to transcribe and translate a human gene! This can be done by transformation.

Genes of interest are first placed into a transformation vector, such as a plasmid. Plasmid vectors are small, circular pieces of DNA that contain genes for antibiotic resistance and restriction sites. Plasmids and the gene of interest are cut with the same restriction enzyme, creating compatible sticky ends. When placed together, the gene is inserted into the plasmid creating recombinant DNA.

The bacteria are then transformed using the recombinant plasmid. In most AP Biology courses, this is done by the heat shock method. Because the plasmid contains a gene for antibiotic resistance, transformed bacteria will be able to grow on a medium that contains antibiotics whereas bacteria without the plasmid will die. This allows scientists to cleverly identify transformed bacteria.

This laboratory technique has not only been used to safely mass-produce important proteins used for medicine, such as insulin; it has an important role in the study of gene expression.

## Human Genome Project

In 1990, an international, publicly funded consortium of scientists was determined to sequence every chromosome, base by base, in the human genome. The latest DNA sequencing machines and innovative computer programming skills were used in this tedious process. A draft of the sequence was first published in 2001. Today, in addition to the human genome, the genomes of many different species are available online for anyone to study at **www.ornl.gov/sci/techresources/ Human_Genome**. These genomes can be used to study our evolutionary history, genetic diseases, and can be used to make pharmaceuticals.

## VIRUSES

Viruses are nonliving agents capable of infecting cells. Why are viruses considered nonliving? They require a host cell's machinery in order to replicate. A virus consists of two main components: a protein capsid and genetic material made of DNA or RNA, depending on the virus.

Viruses vary according to their replication needs as dictated by their host (the cell in which they are replicating) and their type of genome (for example, DNA and RNA). A commonly studied virus is a bacteriophage (a virus that infects bacteria). Bacteriophages undergo two different types of replication cycles, the **lytic cycle** and the **lysogenic cycle**. In the lytic cycle, the lagging strand virus immediately starts using the host cell's machinery to replicate the genetic material and create more protein capsids. These spontaneously assemble into mature viruses and cause the cell to **lyse**, or break open, releasing new viruses into the environment. In the lysogenic cycle, the virus incorporates itself into the host genome and remains dormant until it is triggered to switch into the lytic cycle.

Animal viruses must overcome additional challenges not present in bacterial cells including immune pathways, organelle structures, and different cell machinery. **Retroviruses** like the HIV virus are RNA viruses and use an enzyme called **reverse transcriptase** to convert their RNA genomes into DNA so that they can be inserted into a host genome. RNA viruses have extremely high rates of mutation because they lack error-proofing mechanisms when they replicate their genomes of mutation.

# KEY TERMS

euchromatin

heterochromatin

double helix

nucleotides

five-carbon sugar

phosphate

nitrogenous base

adenine

deoxyribose

guanine

cytosine

thymine

phosphodiester bonds

base pairing

complementary

antiparallel

hydrogen bonds

DNA replication

helicase

replication fork

origins of replication

topoisomerase

DNA polymerase

RNA primase

leading strand

lagging strand

Okazaki fragments

DNA ligase

transcription factors

semiconservative

helicase

polymerase

ligase

topoisomerase

ribose

uracil

messenger RNA (mRNA)

ribosomal RNA (rRNA)

transfer RNA (tRNA)

protein synthesis

transcription

RNA processing

translation

promoters

sense strand

antisense strand

RNA polymerase

exons

introns

spliceosome

poly(A) tail

5' GTP cap

codons

anticodon

A site

P site

E site

stop codons

primary structure

secondary structure

tertiary structure

quaternary structure

chaperon proteins (or chaperonins)

mutation

base substitution

gene rearrangement

nonsense mutation

missense mutation

silent mutation

deletions

duplications

inversions

gene expression

translocations

genetic engineering

restriction enzyme

sticky end

plasmid

transformation

gel electrophoresis

restriction fragment length polymorphism (RFLPs)

DNA fingerprinting

polymerase chain reaction (PCR)

lytic cycle

lysogenic cycle

lyse

retrovirus

reverse transcriptase

# Summary

- DNA is the genetic material of the cell.

- DNA structure is composed of two antiparallel complementary strands, including:
    - deoxyribose sugar and phosphate backbone
    - nitrogenous bases (hydrogen bonded together in the middle)
    - adenine always pairs with thymine, and guanine always pairs with cytosine

- DNA replication provides two copies of the DNA for cell division, which involves the following steps:
    - Helicase begins replication by unwinding DNA and separating the strands at the origin of replication.
    - Topoisomerases reduce supercoiling ahead of the replication fork.
    - RNA primase places an RNA primer down on the template strands.
    - DNA polymerase reads the DNA base pairs and lays down complementary nucleotides to the growing new strand.
    - Replication proceeds continuously on the leading strand.
    - The lagging strand forms with Okazaki fragments, which must be joined later by DNA ligase.

- Transcription involves copying the genetic code directly from the DNA. Transcription factors help initiate RNA polymerase, binding to promoters on the DNA.

- Translation occurs in the cytoplasm as the mRNA carries the message to the ribosome.

- Mutations can result from changes in the DNA message, the mRNA message, or in the interpretation of the codons.

- There are two types of mutations—base substitutions (point mutations) and gene arrangements.

- Some examples of biotechnology are
    - recombinant DNA
    - polymerase chain reaction (PCR)
    - transformation of bacteria
    - the Human Genome Project

# Chapter 8 Drill

Answers and explanations can be found in Chapter 16.

1. A geneticist has discovered a yeast cell, which encodes a DNA polymerase that may add nucleotides in both the 5' to 3' and 3' to 5' directions. Which of the following structures would this cell not likely generate during DNA replication?

   (A) RNA primers
   (B) Okazaki fragments
   (C) Replication fork
   (D) Nicked DNA by topoisomerases

2. A eukaryotic gene, which does not normally undergo splicing, was exposed to benzpyrene, a known carcinogen and mutagen. Following exposure, the protein encoded by the gene was shorter than before exposure. Which of the following types of genetic rearrangements or mutations was likely introduced by the mutagen?

   (A) Silent mutation
   (B) Missense mutation
   (C) Nonsense mutation
   (D) Duplication

3. DNA Replication occurs through a complex series of steps involving several enzymes. Which of the following represents the correct order beginning with the earliest activity of enzymes involved in DNA replication?

   (A) Helicase, ligase, RNA primase, DNA polymerase
   (B) DNA polymerase, RNA primase, helicase, ligase
   (C) RNA primase, DNA polymerase, ligase, helicase
   (D) Helicase, RNA primase, DNA polymerase, ligase

4. If a messenger RNA codon is UAC, which of the following would be the complementary anticodon triplet in the transfer RNA?

   (A) ATG
   (B) AUC
   (C) AUG
   (D) ATT

5. During post-translational modification, the poly-peptide from a eukaryotic cell typically undergoes substantial alteration that results in

   (A) excision of introns
   (B) addition of a poly(A) tail
   (C) formation of peptide bonds
   (D) a change in the overall conformation of a polypeptide

6. Which of the following represents the maximum number of amino acids that could be incorporated into a polypeptide encoded by 21 nucleotides of messenger RNA?

   (A) 3
   (B) 7
   (C) 21
   (D) 42

7. A researcher uses molecular biology techniques to insert a human lysosomal membrane protein into bacterial cells to produce large quantities of this protein for later study. However, only small quantities of this protein result in these cells. What is a possible explanation for this result?

   (A) The membrane protein requires processing in the ER and Golgi, which are missing in the bacterial cells.
   (B) Bacteria do not make membrane proteins.
   (C) Bacteria do not use different transcription factors than humans, so the gene was not expressed.
   (D) Bacteria do not have enough tRNAs to make this protein sequence.

8. *Bam*HI is a restriction enzyme derived from *Bacillus amyloliquefaciens* that recognizes short palindromic sequences in DNA. When the enzyme recognizes these sequences, it cleaves the DNA. What purpose would restriction enzymes have in a bacterium like *Bacillus*?

    (A) They are unnecessary enzymes without a purpose remaining because evolution has produced better enzymes.
    (B) They destroy extra DNA that results from errors in binary fission.
    (C) They protect *Bacillus* from invading DNA due to viruses.
    (D) They prevent, or restrict, DNA replication when the cell isn't ready to copy its DNA.

9. Viruses and bacteria have which of the following in common?

    (A) Ribosomes
    (B) Nucleic acids
    (C) Flagella
    (D) Metabolism

10. Griffith was a researcher that coined the term "transformation" when he noticed that incubating nonpathogenic bacteria with heat-killed pathogenic bacteria produced bacteria that ultimately became pathogenic, or deadly, in mice. What caused the transformation in his experiment?

    (A) DNA from the nonpathogenic bacteria revitalized the pathogenic bacteria.
    (B) Protein from the pathogenic bacteria was taken up by the nonpathogenic bacteria.
    (C) DNA from the pathogenic bacteria was taken up by the nonpathogenic bacteria.
    (D) DNA in the nonpathogenic bacteria turned into pathogenic genes in the absence of pathogenic bacteria.

11. Meselson and Stahl performed an elegant experiment using bacteria grown in heavy nitrogen ($^{15}$N) to show that DNA replication is semi-conservative. After growing the bacteria in heavy nitrogen, they switched the cultures to $^{14}$N for two rounds of replication. They then examined the cells by density centrifugation, separating the components by density. What is a likely result?

    (A) Half of the DNA was labeled with $^{15}$N and half was labeled with $^{14}$N.
    (B) One strand of each new DNA molecule was labeled with $^{15}$N, and the other strand was labeled with $^{14}$N.
    (C) One band averaging the two molecular masses of nitrogen is seen.
    (D) One heavy band and one light band of nitrogen is seen.

12. A biologist systematically removes each of the proteins involved in DNA replication to determine the effect each has on the process. In one experiment, she denatures the DNA to see the results, separating the strands. She sees many short DNA/RNA fragments, as well as some long DNA pieces. What enzyme(s) is/are missing?

    (A) Helicase
    (B) DNA polymerase
    (C) DNA ligase
    (D) RNA primase

## REFLECT

Respond to the following questions:

- For what content topics discussed in this chapter do you feel you have achieved sufficient mastery to answer multiple-choice questions correctly?

- For which content topics discussed in this chapter do you feel you have achieved sufficient mastery to discuss effectively in a free response?

- For which content topics discussed in this chapter do you feel you need more work before you can answer multiple-choice questions correctly?

- For which content topics discussed in this chapter do you feel you need more work before you can discuss effectively in a free response?

- What parts of this chapter are you going to re-review?

- Will you seek further help, outside this book (such as a teacher, tutor, or AP Central), on any of the content in this chapter—and if so, on what content?

# Chapter 9
# Cell Reproduction

## CELL DIVISION

Every second, thousands of cells are dying throughout our bodies. Fortunately, the body replaces them at an amazing rate. In fact, **epidermal**, or skin, cells die off and are replaced so quickly that the average 18-year-old grows an entirely new skin every few weeks. The body keeps up this unbelievable rate thanks to the mechanisms of **cell division**.

This chapter takes a closer look at how cells divide. But remember, cell division is only a small part of the life cycle of a cell. Most of the time, cells are busy carrying out their regular activities. Since we covered DNA replication in Chapter 8, let's now look at how cells pass their genetic material to their offspring.

## THE CELL CYCLE

Every cell has a life cycle—the period from the beginning of one division to the beginning of the next. The cell's life cycle is known as the **cell cycle**. The cell cycle is divided into two periods: **interphase** and **mitosis**. Take a look at the cell cycle of a typical cell:

Notice that most of the life of a cell is spent in interphase.

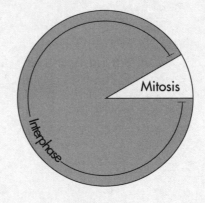

# INTERPHASE: THE GROWING PHASE

Interphase is the time span from one cell division to another. We call this stage interphase (*inter-* means between) because the cell has not yet started to divide. Although biologists sometimes refer to interphase as the "resting stage," the cell is definitely not inactive. This phase is when the cell carries out its regular activities. All the proteins and enzymes it needs to grow are produced during interphase.

## The Three Stages of Interphase

Interphase can be divided into three stages: $G_1$, $G_2$, and **S phase.**

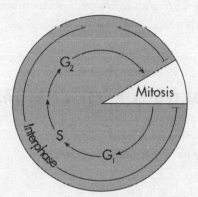

The most important phase is the S phase. That's when the cell replicates its genetic material. The first thing a cell has to do before undergoing mitosis is to duplicate all of its chromosomes, which contain the organism's DNA "blueprint." During interphase, every single chromosome in the nucleus is duplicated.

You'll notice that the original chromosome and its duplicate are still linked, like Siamese twins. These identical chromosomes are now called **sister chromatids** (each individual structure is called a chromatid). The chromatids are held together by a structure called the **centromere**. Although the chromosomes have been duplicated, they are still considered a single unit. Once duplication has been done, we're ready for the big breakup: mitosis.

We've already said that replication occurs during the S phase of interphase, so what happens during $G_1$ and $G_2$? During these stages, the cell produces proteins and enzymes. For example, during G1 the cell produces all of the enzymes required for DNA replication (as we saw in Chapter 8, that means DNA helicase, DNA polymerase, and DNA ligase). By the way, "G" stands for "gap," but we can also associate it with "growth." These three phases are highly regulated by checkpoints and special proteins called cyclins and cyclin-dependent kinases (CDKs).

Let's recap:

- The cell cycle consists of two things: interphase and mitosis.

- During the S phase of interphase, the chromosomes replicate.

- Growth and preparation for mitosis occur during the $G_1$ and $G_2$ stages of interphase.

## MITOSIS: THE DANCE OF THE CHROMOSOMES

Once the chromosomes have replicated, the cell is ready to begin mitosis. Mitosis is the period when the cell divides. Mitosis consists of a sequence of four stages: **prophase**, **metaphase**, **anaphase**, and **telophase**.

## Stage 1: Prophase

One of the first signs of prophase is the disappearance of the nucleolus. In prophase, the chromosomes thicken, forming coils upon coils, and become visible. (During interphase, the chromosomes are not visible. Rather, the genetic material is scattered throughout the nucleus and is called **chromatin**. It is only during prophase that we can properly speak about the chromosomes.)

Now the cell has plenty of room to "sort out" the chromosomes. Remember centrioles? During prophase, these cylindrical bodies found within microtubule organizing centers (MTOCs) start to move away from each other, toward opposite ends of the cell. The centrioles will spin out a system of microtubules known as the **spindle fibers**. These spindle fibers will attach to a structure on each chromatid called a **kinetochore**. The kinetochores are part of the centromere.

## Stage 2: Metaphase

The next stage is called metaphase. The chromosomes now begin to line up along the equatorial plane, or the **metaphase plate**, of the cell. That's because the spindle fibers are attached to the kinetochore of each chromatid.

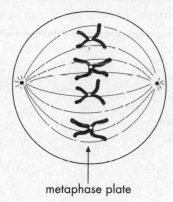

metaphase plate

## Stage 3: Anaphase

During anaphase, the sister chromatids of each chromosome separate at the centromere and migrate to opposite poles. The chromatids are pulled apart by the microtubules, which begin to shorten. Each half of a pair of sister chromatids now moves to opposite poles of the cell. Non-kinetochore microtubules elongate the cell.

## Stage 4: Telophase

The final phase of mitosis is telophase. A nuclear membrane forms around each set of chromosomes and the nucleoli reappear.

The nuclear membrane is ready to divide. Now it's time to split the cytoplasm in a process known as **cytokinesis**. Look at the figure below and you'll notice that the cell has begun to split along a **cleavage furrow** (which is produced by actin microfilaments):

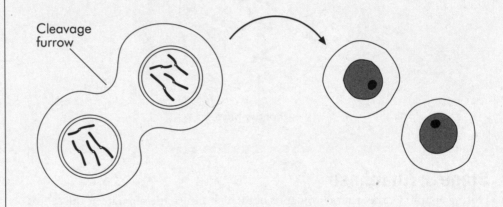

Cleavage furrow

A cell membrane forms about each cell and they split into two distinct daughter cells. The division of the cytoplasm yields two daughter cells.

Here's one thing to remember: Cytokinesis occurs differently in plant cells. The cell doesn't form a cleavage furrow. Instead, a partition called a **cell plate** forms down the middle region.

## Stage 5: Interphase

Once the daughter cells are produced, they reenter the initial phase—interphase—and the whole process starts over. The cell goes back to its original state. Once again, the chromosomes become invisible, and the genetic material is called chromatin.

# But How Will I Remember All That?

For mitosis, you may already have your own mnemonic. If not, here's a table with a mnemonic we created for you.

| IPMAT | |
|---|---|
| Interphase | **I** is for **Interlude** |
| Prophase | **P** is for **Prepare** |
| Metaphase | **M** is for **Meet** |
| Anaphase | **A** is for **Apart** |
| Telophase | **T** is for **Tear** |

## Purpose of Mitosis

Mitosis has two purposes:

- to produce daughter cells that are identical copies of the parent cell
- to maintain the proper number of chromosomes from generation to generation

For our purposes, we can say that mitosis occurs in just about every cell except sex cells. When you think of mitosis, remember: "Like begets like." Hair cells "beget" other hair cells, skin cells "beget" other skin cells, and so on. Mitosis is involved in growth, repair, and asexual reproduction.

# HAPLOIDS VERSUS DIPLOIDS

Every organism has a certain number of chromosomes. For example, fruit flies have 8 chromosomes, humans have 46 chromosomes, and dogs have 78 chromosomes. It turns out that most eukaryotic cells in fact have two full sets of chromosomes—one set from each parent. Humans, for example, have two sets of 23 chromosomes, giving us our grand total of 46.

A cell that has both sets of chromosomes is a **diploid cell**, and the zygotic chromosome number is given as "$2n$." That means we have two copies of each chromosome. In humans, for example, the diploid number of chromosomes is 46.

If a cell has only one set of chromosomes, we call it a **haploid cell**. This kind of cell is given the symbol "$n$." For example, we would say that the haploid number of chromosomes for humans is 23.

Remember:

- *Diploid* refers to any cell that has two sets of chromosomes.
- *Haploid* refers to any cell that has one set of chromosomes.

Why do we need to know the terms *haploid* and *diploid*? Because they are extremely important when it comes to sexual reproduction. As we've seen, 46 is the normal diploid number for human beings. We can say, therefore, that human cells have 46 chromosomes. However, this isn't entirely correct.

Human chromosomes come in pairs called **homologues**. So while there are 46 of them altogether, there are actually only 23 *distinct* chromosomes. The **homologous chromosomes** which make up each pair are similar in size and shape and express similar traits. This is the case in all sexually reproducing organisms. In fact, this is the essence of sexual reproduction: Each parent donates half its chromosomes to its offspring.

## Gametes

Although most cells in the human body are diploid (filled with pairs of chromosomes), there are special cells that are haploid (unpaired). These haploid cells are called **sex cells**, or **gametes**. Why do we have haploid cells?

As we've said, an offspring has one set of chromosomes from each of its parents. A parent, therefore, contributes a gamete with one set that will be paired with the set from the other parent to produce a new diploid cell, or zygote. This produces offspring that are genetically different from either parent.

## AN OVERVIEW OF MEIOSIS

To preserve the diploid number of chromosomes in an organism, each parent must contribute only half of its chromosomes. This is the point of **meiosis**. Meiosis is the production of gametes. Since sexually reproducing organisms need only haploid cells for reproduction, meiosis is limited to sex cells in special sex organs called **gonads**. In males, the gonads are the **testes**, while in females they are the **ovaries**. The special cells in these organs—also known as **germ cells**—produce haploid cells ($n$), which then combine to restore the diploid ($2n$) number during fertilization:

$$\text{female gamete } (n) + \text{male gamete } (n) = \text{zygote } (2n)$$

When it comes to genetic variation, meiosis is a big plus. Variation, in fact, is the driving force of evolution. The more variation there is in a population, the more likely it is that some members of the population will survive extreme changes in the environment. Meiosis is far more likely to produce these sorts of variations than is mitosis, and therefore confers selective advantage on sexually reproducing organisms. We'll come back to this theme in Chapter 11.

## A Closer Look at Meiosis

Meiosis actually involves two rounds of cell division: **meiosis I** and **meiosis II**.

Before meiosis begins, the diploid cell goes through interphase. Just as in mitosis, double-stranded chromosomes are formed during this phase.

## Meiosis I

Meiosis I consists of four stages: prophase I, metaphase I, anaphase I, and telophase I.

### Prophase I

Prophase I is a little more complicated than regular prophase. As in mitosis, the nuclear membrane disappears, the chromosomes become visible, and the centrioles move to opposite poles of the nucleus. But that's where the similarity ends.

The major difference involves the movement of the chromosomes. In meiosis, the chromosomes line up side-by-side with their counterparts (homologues). This event is known as **synapsis**.

Synapsis involves two sets of chromosomes that come together to form a **tetrad** (or a **bivalent**). A tetrad consists of four chromatids. Synapsis is followed by **crossing-over**, the exchange of segments between homologous chromosomes.

What's unique in prophase I is that "pieces" of chromosomes are exchanged between the homologous partners. This is one of the ways organisms produce genetic variation. By the end of prophase I, the chromosomes will have exchanged regions containing several **alleles**, or different forms of the same gene. By the end of prophase, the homologous chromosomes are held together only at specialized regions called **chiasmata**.

## Metaphase I

As in mitosis, the chromosome pairs—now called tetrads—line up at the metaphase plate. By contrast, you'll recall that in regular metaphase the chromosomes lined up individually. One important concept to note is that the alignment during metaphase is random, so which copy of each chromosome that ends up in a daughter cell is random.

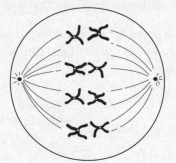

## Anaphase I

During anaphase I, one of each pair of chromosomes within a tetrad separates and moves to opposite poles. Notice that the chromosomes do not separate at the centromere. They separate with their centromeres intact.

The chromosomes now go on to their respective poles.

## Telophase I

During telophase I, the nuclear membrane forms around each set of chromosomes.

Finally, the cells undergo cytokinesis, leaving us with two daughter cells. Notice that at this point the nucleus contains the haploid number of chromosomes, but each chromosome is a duplicated chromosome.

## Meiosis II

The purpose of the second meiotic division is to separate the duplicated chromosomes, and it is virtually identical to mitosis. Let's run through the steps in meiosis II.

After a brief period, the cell undergoes a second round of cell division. During prophase II, the chromosomes once again condense and become visible. In metaphase II, the chromosomes move toward the metaphase plate. This time they line up single file, not as pairs. During anaphase II, the chromatids of each chromosome split at the centromere and are pulled to opposite ends of the cell. At telophase II, a nuclear membrane forms around each set of chromosomes and a total of *four* haploid cells are produced:

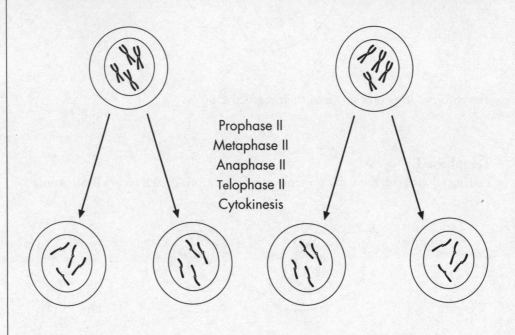

Prophase II
Metaphase II
Anaphase II
Telophase II
Cytokinesis

## Gametogenesis

Meiosis is also known as **gametogenesis.** If sperm cells are produced then meiosis is called **spermatogenesis.** During spermatogenesis, four sperm cells are produced for each diploid cell. If an egg cell or an ovum is produced, this process is called **oogenesis.**

Oogenesis is a little different from spermatogenesis. Oogenesis produces only one ovum, not four. The other three cells, called **polar bodie**s, get only a tiny amount of cytoplasm and eventually degenerate. Why does oogenesis produce only one ovum? Because the female wants to conserve as much cytoplasm as possible for the surviving gamete, the **ovum.**

Here's a summary of the major differences between mitosis and meiosis:

| MITOSIS | MEIOSIS |
|---|---|
| • Occurs in somatic (body) cells | • Occurs in germ (sex) cells |
| • Produces identical cells | • Produces gametes |
| • Diploid cell → diploid cells | • Diploid cell → haploid cells |
| • 1 cell becomes 2 cells | • 1 cell becomes 4 cells |
| • Number of divisions: 1 | • Number of divisions: 2 |

## Mutations

Sometimes, a set of chromosomes has an extra or a missing chromosome. This occurs because of **nondisjunction**—the chromosomes failed to separate properly during meiosis. This error, which produces the wrong number of chromosomes in a cell, results in severe genetic defects. For example, humans typically have 23 pairs of chromosomes, but individuals with **Down syndrome** have three—instead of two—copies of the 21st chromosome.

Chromosomal abnormalities also occur if one or more segments of a chromosome break. The most common example is translocation (a segment of a chromosome moves to another chromosome).

Here's an example of a translocation:

A    B
Chromosomes

A    B
Chromosomes

**Translocation** is a mutation that occurs during meiosis when a chromosome breaks and the fragment attaches to a nonhomologous chromosome.

Fortunately, in most cases, damaged DNA can usually be repaired with special repair enzymes.

# Gene Regulation

What controls gene transcription, and how does an organism express only certain genes? Most of what we know about gene regulation comes from our studies of *E. coli*. In bacteria, the region of bacterial DNA that regulates gene expression is called an **operon**. One of the best-understood operons is the *lac* operon, which controls expression of the enzymes that break down lactose.

The operon consists of four major parts: structural genes, the regulatory gene, the promoter gene, and the operator.

- Structural genes are genes that code for enzymes needed in a chemical reaction. These genes will be transcribed at the same time to produce particular enzymes. In the *lac* operon, three enzymes (beta galactosidase, galactose permease, and thiogalactoside transacetylase) involved in digesting lactose are coded for.

- The **promoter gene** is the region where the RNA polymerase binds to begin transcription.

- The operator is a region that controls whether transcription will occur.

- The regulatory gene codes for a specific regulatory protein called the repressor. The repressor is capable of attaching to the operator and blocking transcription. If the repressor binds to the operator, transcription will not occur. On the other hand, if the repressor does not bind to the operator, RNA polymerase moves right along the operator and transcription occurs. In the *lac* operon, the **inducer**, lactose, binds to the repressor, causing it to fall off the **operator**, and "turns on" transcription.

Other operons, such as the *trp* operon, operate in a similar manner except that this mechanism is continually "turned on" and is only "turned off" in the presence of high levels of the amino acid, tryptophan. Tryptophan is a product of the pathway that codes for the *trp* operon. When tryptophan combines with the *trp* repressor protein, it causes the repressor to bind to the operator, which turns the operon "off" thereby blocking transcription. In other words, a high level of tryptophan acts to repress the further synthesis of tryptophan.

The following diagram shows the structure of a gene.

The next two diagrams illustrate the *lac* operon (inducer) and the *trp* operon (repressor).

This diagram illustrates a *lac* (inducer) operon.

This diagram represents the
*trp* (repressor) operon.

## KEY TERMS

epidermal
cell division
cell cycle
interphase
mitosis
$G_1$ phase
$G_2$ phase
S phase
sister chromatids
centromere
prophase
metaphase
anaphase
telophase
chromatin
spindle fibers
kinetochores
metaphase plate
cytokinesis
cleavage furrow
cell plate
diploid cell
haploid cell
homologues
homologous chromosomes
sex cells (or gametes)
meiosis

gonads
testes
ovaries
germ cells
meiosis I
meiosis II
synapsis
tetrad (or bivalent)
crossing-over
alleles
chiasmata
gametogenesis
spermatogenesis
oogenesis
polar bodies
ovum
nondisjunction
Down syndrome
translocation
operon
structural genes
promoter gene
operator
regulatory gene
inducer

# Summary

- The cell cycle is divided into interphase and mitosis, or cellular division.

- The three stages of interphase are $G_1$, $G_2$, and S phase.
  - S phase is the "synthesis" phase, when chromosomes replicate.
  - Growth and preparation for mitosis occur in $G_1$ and $G_2$.

- Mitosis is cellular division and occurs in four stages: prophase, metaphase, anaphase, and telophase.
  - Prophase is when the nuclear envelope disappears and chromosomes condense.
  - Next is metaphase, when chromosomes align at the metaphase plate and mitotic spindles attach to kinetochores.
  - Anaphase pulls the chromosomes away from the center.
  - Telophase terminates mitosis, and the two new nuclei form.
  - The process of cytokinesis, which occurs during telophse, ends mitosis, as the cytoplasm and plasma membranes pinch to form two distinct, identical daughter cells.

- Meiosis produces four genetically distinct haploid gametes. Meiosis involves two rounds of cell division: meiosis I and meiosis II.
  - Meiosis I is when the homologous chromosome pairs separate.
  - Meiosis II is when the sister chromatids separate.
  - Mutations, such as nondisjunction events or whole chromosome translocations, can occur as a result of crossing-over.

- Gene expression is tightly regulated.
  - Examples of gene regulation in bacteria are *lac* operons and *trp* operons.
  - In eukaryotes, gene expression is controlled in the following ways:
    - DNA is compacted into chromatin.
    - Transcription is controlled by transcription factors.
    - RNA processing, RNA regulation, and RNA interference control RNA duration in the cytoplasm.
    - Translation is regulated by translation initiation factors.
    - Proteins are ultimately controlled by post-translational modifications, localization, and degradation.

# Chapter 9 Drill

Answers and explanations can be found in Chapter 16.

1. A scientist is testing new chemicals designed to stop the cell cycle at various stages of mitosis. Upon applying one of the chemicals, she notices that all of the cells appear as shown below. Which of the following best explains how the chemical is likely acting on the cells?

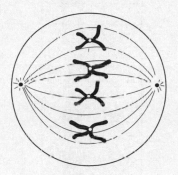

(A) The chemical has arrested the cells in prophase and has prevented attachment of the spindle fibers to the kinetochore.

(B) The chemical has arrested the cells in metaphase and has prevented dissociation of the spindle fibers from the kinetochore.

(C) The chemical has arrested the cells in metaphase and is preventing the shortening of the spindle fibers.

(D) The chemical has arrested the cells in anaphase and is preventing the formation of a cleavage furrow.

2. A unique lineage of cells was identified which undergoes one round of meiosis, but it results in four daughter cells without exchange of alleles. It is suspected that there is a defect during the crossing-over event. During which phase of meiosis is the defect likely occurring?

(A) Prophase I
(B) Metaphase I
(C) Telophase I
(D) Prophase II

3. In the *lac* operon, lactose binds to a molecule, which blocks RNA polymerase activity, causing it to dissociate and permit transcription. Which of the following terms most accurately describes the role of lactose and its molecule target?

(A) Lactose is a repressor, and its target molecule is an inducer.

(B) Lactose is a promoter, and its target molecule is an inducer.

(C) Lactose is an inducer, and its target molecule is a repressor.

(D) Lactose is a repressor, and its target molecule is a promoter.

Questions 4–5 refer to the following graph and paragraph.

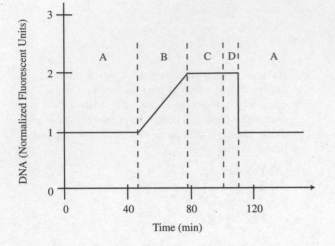

An experiment is performed to evaluate the amount of DNA present during a complete cell cycle. All of the cells were synced prior to the start of the experiment. During the experiment, a fluorescent chemical was applied to cells, which would only fluoresce when bound to DNA. The results of the experiment are shown above. Differences in cell appearance by microscopy or changes in detected DNA were determined to be phases of the cell cycle and are labeled with the letters A–D.

4. Approximately how long does S phase take to occur in these cells?

(A) 15 min
(B) 20 min
(C) 30 min
(D) 40 min

5. During which of the labeled phases of the experiment would the cell undergo anaphase?

(A) Phase A
(B) Phase B
(C) Phase C
(D) Phase D

Question 6 represents a question requiring a numeric answer. Calculate the correct answer for each question.

6. A new mammal has recently been discovered in the Amazonian jungle. A karyotype was performed on gametic cells and revealed that the animal had 13 different chromosomes. How many chromosomes would you expect to find in a diploid cell of the organism following completion of S phase? Give your answer to the nearest whole number.

7. Trisomy 21, which results in Down syndrome, results from nondisjunction of chromosome 21 in humans. Nondisjunction occurs when two homologous chromosomes, or two sister chromatids, do not separate. Which of the following describes the mechanism of this defect?

(A) During DNA replication in S phase of the cell cycle, the two new strands do not separate.
(B) During mitosis, at the metaphase plate, nonsister chromatids do not separate.
(C) The mitotic spindle attaches to chiasmata rather than kinetochores.
(D) Recombination during crossing-over remains incomplete.

8. A researcher treated dividing cells with colchicine, which disrupts the mitotic spindle. When he then observed the cells, what phase of the cell cycle were the cells stalled in?

   (A) Interphase
   (B) Telophase
   (C) Metaphase
   (D) Anaphase

9. A researcher isolates DNA from different types of cells and determines the amount of DNA for each type of cell. In order of increasing DNA content, which of the following would have the least amount of DNA to the greatest amount of DNA?

   (A) Sperm, neuron, liver
   (B) Ovum, muscle, taste bud
   (C) Liver, heart, intestine
   (D) Skin, neuron, ovum

10. Which of the following mutations would be least likely to have any discernable phenotype on the individual?

   (A) Translocation of the last 1,000 base pairs of chromosome 1 onto chromosome 2
   (B) Nondisjunction of chromosome 7 to produce trisomy-7
   (C) Nondisjunction of chromosome 8 to produce monosomy-8
   (D) Deletion of 400 base pairs, resulting in the loss of an enhancer

# REFLECT

Respond to the following questions:

- For what content topics discussed in this chapter do you feel you have achieved sufficient mastery to answer multiple-choice questions correctly?

- For which content topics discussed in this chapter do you feel you have achieved sufficient mastery to discuss effectively in a free response?

- For which content topics discussed in this chapter do you feel you need more work before you can answer multiple-choice questions correctly?

- For which content topics discussed in this chapter do you feel you need more work before you can discuss effectively in a free response?

- What parts of this chapter are you going to re-review?

- Will you seek further help, outside this book (such as a teacher, tutor, or AP Central), on any of the content in this chapter—and if so, on what content?

# Chapter 10
# Heredity

# GREGOR MENDEL: THE FATHER OF GENETICS

What is genetics? In its simplest form, genetics is the study of heredity. It explains how certain characteristics are passed on from parents to children. Much of what we know about genetics was discovered by the monk **Gregor Mendel** in the 19th century. Since then, the field of genetics has vastly expanded. As scientists study the mechanisms of genetics, they've developed new ways of manipulating genes. For example, scientists have isolated the gene that makes insulin, a human hormone, and now use bacteria to make large quantities of it. But before we get ahead of ourselves, let's study the basic rules of genetics.

Let's begin then with some of the fundamental points of genetics:

- Every **trait**—or expressed characteristic—is produced by hereditary factors known as **genes**. A gene is a segment of a chromosome. Within a chromosome, there are many genes, each controlling the inheritance of a particular trait. For example, in pea plants, there's a gene on the chromosome that codes for seed coat. The position of a gene on a chromosome is called a **locus**.

- Diploid organisms (organisms that have two sets of chromosomes) usually have two copies of a gene, one on each homologous chromosome. These copies may be different from one another—that is, they may be **alleles**, or alternate forms of the same gene. For example, if we're talking about the height of a pea plant, there's an allele for tall and an allele for short. In other words, both alleles are alternate forms of the gene for height.

- An allele can be **dominant** or **recessive**. In simple cases, an organism can express contrasting conditions. For example, a plant can be tall or short. The convention is to assign one of two letters for the two different alleles. The dominant allele receives a capital letter and the recessive allele receives a lowercase of the same letter. For instance, we might give the dominant allele for height in pea plants a "T" for tall. This means that the recessive allele would be "t."

- When an organism has two identical alleles for a given trait, the organism is **homozygous**. For instance, TT and tt both represent homozygous organisms. TT is homozygous dominant and tt is homozygous recessive. If an organism has two *different* alleles for a given trait, Tt, the organism is **heterozygous**.

- When discussing the physical appearance of an organism, we refer to its **phenotype**. The phenotype tells us what the organism looks like. When talking about the genetic makeup of an organism, we refer to its **genotype**. The genotype tells us which alleles the organism possesses.

One of the major ways ETS likes to test genetic information is by having you do crosses. Crosses involve the mating of hypothetical organisms with specific phenotypes and genotypes. We'll look at some examples in a moment, but for now, keep these test-taking tips in mind:

- Label each generation in the cross. The first generation in an experiment is always called the **parent**, or **P1 generation**. The offspring of the P1 generation are called the **filial**, or **F1 generation**. The next generation, the grandchildren, is called the **F2 generation**.

- Always write down the symbols you're using for the alleles, along with a key to remind yourself what the symbols refer to. Use uppercase for dominant alleles and lowercase for recessive alleles.

Now let's look at some basic genetic principles.

## MENDELIAN GENETICS

One of Mendel's hobbies was to study the effects of cross-breeding on different strains of pea plants. Mendel worked exclusively with true-breeder pea plants. This means the plants he used were genetically pure and consistently produced the same traits. For example, tall plants always produced tall plants; short plants always produced short plants. Through his work, he came up with three principles of genetics: the **law of dominance**, the **law of segregation**, and the **law of independent assortment**.

## The Law of Dominance

Mendel crossed two true-breeding plants with contrasting traits: tall pea plants and short pea plants. This type of cross is called a **monohybrid cross**, which means that only one trait is being studied. In this case, the trait was height.

To his surprise, when Mendel mated these plants, the characteristics didn't blend to produce plants of average height. Instead, all the offspring were tall.

Genotype ⟶ T T

Phenotype ⟶ Tall plant

Short plant

×

Tall plant

Mendel recognized that one trait must be masking the effect of the other trait. This is called the law of dominance. The dominant, tall allele, T, somehow masked the presence of the recessive, short allele, t. Consequently, all a plant needs is one tall allele to make it tall.

## Monohybrid Cross

A simple way to represent a cross is to set up a **Punnett square**. Punnett squares are used to predict the results of a cross. Let's construct a Punnett square for the cross between Mendel's tall and short pea plants. Let's first designate the alleles for each plant. As we saw earlier, we can use the letter "T" for the tall, dominant allele and "t" for the short, recessive allele.

Since one parent was a pure, tall pea plant, we'll give it two dominant alleles (TT homozygous dominant). The other parent was a pure, short pea plant, so we'll give it two recessive alleles (tt homozygous recessive). Let's put the alleles for one of the parents across the top of the box and the alleles for the other parent along the side of the box.

Now we can fill in the four boxes by matching the letters. What are the results for the F1 generation?

Each offspring received one allele from each parent. They all received one T and one t. They're all Tt! Our parents had duplicate copies of single alleles, TT and tt, respectively. We could therefore refer to them as homozygous. The offspring, on the other hand, are heterozygous: They possess one copy of each allele.

Let's compare the results of this cross with what we already know about meiosis. From meiosis, we know that when gametes are formed, the chromosomes separate so that each cell gets one copy of each chromosome. We now know that chromosomes are made up of genes, and genes consist of alleles. We've just seen that alleles also separate and recombine. We can say, therefore, that each allele in a Punnett square also "represents" a gamete:

When fertilization occurs, chromosomes—along with the alleles they carry—get paired up in a new combination.

## The Law of Segregation

Next, Mendel took the offspring and self-pollinated them. Let's use a Punnett square to spell out the results. This time we're starting with the offspring of the first generation—F1. Take a look at the results:

```
        T    t
    T | TT | Tt |
    t | Tt | tt |
```

F2 Generation

One of the offspring could be a short pea plant! The short-stemmed trait reappeared in the F2 generation. How could that happen? Once again, the alleles separated and recombined, producing a new combination for this offspring. The cross resulted in one offspring with a pair of recessive alleles, tt. Because there is no T (dominant) allele around to mask the expression of the short, recessive allele, our new plant could wind up short.

Although all of the F1 plants appear to be tall, the alleles separate and recombine during the cross. This is an example of the law of segregation.

What about the genotype and phenotype for this cross? Remember, genotype refers to the genetic makeup of an organism, whereas phenotype refers to the appearance of the organism. Using the results of our Punnett square, what is the ratio of phenotypes and genotypes in the offspring?

Let's sum up the results. We have four offspring with two different phenotypes: three of the offspring are tall, whereas one of them is short. On the other hand, we have three genotypes: 1 TT, 2 Tt, and 1 tt.

Here's a summary of the results:

- The ratio of phenotypes is 3:1 (three tall:one short).
- The ratio of genotypes is 1:2:1 (one TT:two Tt:one tt).

## The Law of Independent Assortment

So far, we have looked at only one trait: tall versus short. What happens when we study two traits at the same time? The two traits also segregate randomly. This is an example of independent assortment. For example, let's look at two traits in pea plants: height and color. When it comes to height, a pea plant can be either tall or short. As for color, the plant can be either green or yellow, with green being dominant. This gives us four alleles. By the law of independent assortment, these four alleles can combine to give us four different gametes:

TG  Tg  tG  tg

## Dihybrid Cross

Keep in mind that the uppercase letter refers to the dominant allele. Therefore, "T" refers to tall and "G" refers to green, whereas "t" refers to short and "g" refers to yellow. Now let's set up a cross between plants differing in two characteristics—called a **dihybrid cross**—using these four gametes and see what happens.

Each trait will act independently, meaning that a plant that is tall can be either green or yellow. Similarly, a green plant can be either tall or short.

Here is the Punnett Square for a cross between two double heterozygotes (Tt Gg):

|      | TG   | Tg   | tG   | tg   |
|------|------|------|------|------|
| TG   | TTGG | TTGg | TtGG | TtGg |
| Tg   | TTGg | TTgg | TtGg | Ttgg |
| tG   | TtGG | TtGg | ttGG | ttGg |
| tg   | TtGg | Ttgg | ttGg | ttgg |

This is an example of the law of independent assortment. Each of the traits segregated independently. Don't worry about the different combinations in the cross—you'll make yourself dizzy with all those letters. Simply memorize the phenotype ratio of the pea plants. For the 16 offspring there are

- 9 tall and green

- 3 tall and yellow

- 3 short and green

- 1 short and yellow

That's 9:3:3:1. Since Mendel's laws hold true for most of the traits they'll ask you about on the AP test, simply learning the ratios of offspring for this type of cross will help you nail any questions that come up.

The Punnett square method works well for monohybrid crosses and helps us visualize the possible combinations. However, a better method for predicting the likelihood of certain results from a dihybrid cross is to apply the law of probability. For dihybrid ratios, the law states that the probability that two or more independent events will occur simultaneously is equal to the product of the probability that each will occur independently. To illustrate the product rule, let's consider again the cross between two identical dihybrid tall, green plants with the genotype TtGg. To find the probability of having a tall, yellow plant, simply multiply the probabilities of each event. If the probability of being tall is $\frac{3}{4}$ and the probability of being yellow is $\frac{1}{4}$, then the probability of being tall and yellow is $\frac{3}{4} \times \frac{1}{4} = \frac{3}{16}$.

Let's summarize Mendel's three laws.

| SUMMARY OF MENDEL'S LAWS | |
|---|---|
| **Laws** | **Definition** |
| Law of Dominance | One trait masks the effects of another trait. |
| Law of Segregation | Alleles can segregate and recombine. |
| Law of Independent Assortment | Traits can segregate and recombine independently of other traits. |

## Test Cross

Suppose we want to know if a tall plant is homozygous (TT) or heterozygous (Tt). Its physical appearance doesn't necessarily tell us about its genetic makeup. The only way to determine its genotype is to cross the plant with a recessive, short plant, tt. This is known as a **test cross**, or back cross. Using the recessive plant, there are only two possibilities: (1) TT × tt or (2) Tt × tt. Let's take a look.

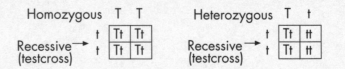

If none of the offspring is short, our original plant must have been homozygous, TT. If, however, even one short plant appears in the bunch, we know that our original pea plant was heterozygous, Tt. In other words, it wasn't a pure-breeding plant. A test cross uses a recessive organism to determine the genotype of an organism of unknown genotype.

# BEYOND MENDELIAN GENETICS

Not all patterns of inheritance obey the principles of Mendelian genetics. In fact, many traits we observe are due to a combined expression of alleles. Here are a couple of examples of non-Mendelian forms of inheritance:

- **Incomplete dominance (blending inheritance):** In some cases, the traits will blend. For example, if you cross a white snapdragon plant (genotype WW) with a red snapdragon plant (RR), the resulting progeny will be pink (RW).

- **Codominance:** Sometimes you'll see an equal expression of both alleles. For example, an individual can have an AB blood type. In this case, each allele is equally expressed. That is, both the A allele and the B allele are expressed ($I^A I^B$). That's why the person is said to have AB blood.

- **Polygenic inheritance:** In some cases, a trait results from the interaction of *many* genes. Each gene will have a small effect on a particular trait. Height, skin color, and weight are all examples of polygenic traits.

- **Multiple alleles**: Some traits are the product of many different alleles that occupy a specific gene locus. The best example is the ABO blood group system in which three alleles ($I^A$, $I^B$, and *i*) determine blood type.

- **Linked genes:** Sometimes genes on the *same* chromosome stay together during assortment and move as a group. The group of genes is considered linked and tends to be inherited together. For example, the genes for flower color and pollen shape are linked on the same chromosomes and show up together. This pattern has led to methods for mapping human chromosomes.

Since linked genes are found on the same chromosome, they cannot segregate independently. That means they do not follow the probability rule and the expected results from a dihybrid cross.

In addition, the frequency of crossing-over between any two linked alleles is proportional to the distance between them. That is, the farther apart two linked alleles are on a chromosome, the more often the chromosome will break between them. This finding led to recombination mapping—mapping of linkage groups with each map unit being equal to 1 percent recombination. For example, if two linked genes, A and B, recombine with a frequency of 15 percent, and B and C recombine with a frequency of 9 percent, and A and C recombine with a frequency of 24 percent, what is the sequence and the distance between them?

The sequence and the distance of A-B-C is

15 units                    9 units
A_____B_____C

If the recombination frequency between A and C had been 6 percent instead of 24 percent, the sequence and distance of A-B-C would instead be

6 units        9 units
A_____C_____B

## SEX-LINKED TRAITS

We already know that humans contain 23 pairs of chromosomes. Twenty-two of the pairs of chromosomes are called **autosomes**. They code for many different traits. The other pair contains the **sex chromosomes**. This pair determines the sex of an individual. A female has two X chromosomes. A male has one X and one Y chromosome—an X from his mother and a Y from his father. Some traits, such as **color blindness** and **hemophilia**, are carried on sex chromosomes. These are called **sex-linked traits**. Most sex-linked traits are found on the X chromosome and are more properly referred to as "X-linked."

Since males have one X and one Y chromosome, what happens if a male has a defective X chromosome? Unfortunately, he'll express the sex-linked trait. Why? Because his one and only X chromosome is defective. He doesn't have another X to mask the effect of the bad X. However, if a female has only one defective X chromosome, she won't express the sex-linked trait. For her to express the trait, she has to inherit two defective X chromosomes. A female with one defective X is called a **carrier**. Although she does not exhibit the trait, she can still pass it on to her children.

You can also use the Punnett square to figure out the results of sex-linked traits. Here's a classic example: A male who has normal color vision and a woman who is a carrier for color blindness have children. How many of the children will be color-blind? To figure out the answer, let's set up a Punnett square:

X̄ = defective X

Notice that we placed a bar above any defective X to indicate the presence of a defective allele. And now for the results. The couple would have one son who is color-blind, a normal son, a daughter who is a carrier, and a normal daughter. The color-blind child is a son.

## Barr Bodies

A look at the cell nucleus of normal females will reveal a dark-staining body known as a **Barr body**. A Barr body is an X chromosome that is condensed and visible. In every female cell, one X chromosome is activated and the other X chromosome is deactivated during embryonic development. Surprisingly, the X chromosome destined to be inactivated is randomly chosen in each cell. Therefore, in every tissue in the adult female one X chromosome remains condensed and inactive. However, this X chromosome is replicated and passed on to a daughter cell.

## Other Sex-Linked Diseases

- Duchenne muscular dystrophy
- Vitamin D resistant rickets
- Hereditary nephritis
- Juvenile gout

# KEY TERMS

Gregor Mendel

trait

genes

locus

alleles

dominant

recessive

homozygous

heterozygous

phenotype

genotype

parent, or P1 generation

filial, or F1 generation

F2 generation

law of dominance

law of segregation

law of independent assortment

monohybrid cross

Punnett square

dihybrid cross

test cross

incomplete dominance

codominance

polygenetic inheritance

multiple alleles

linked genes

autosomes

sex chromosomes

color blindness

hemophilia

sex-linked traits

carrier

Barr body

# Summary

- Genetics is the study of heredity. The foundation of genetics is that every trait is produced by genes, which are found on chromosomes.

- Diploid organisms carry two alleles for every gene, one on each homologous chromosome.
  - Alleles can be either dominant or recessive. When two alleles are the same, the organism is homozygous for that trait. When the two alleles are different, the organism is heterozygous for the trait.

- The genes are described as the genotype of the organism. The physical expression of the traits, or what you see in the organism, is the phenotype.

- In Mendelian genetics, there are three important laws: The Law of Dominance, the Law of Segregation, and the Law of Independent Assortment.

- Non-Mendelian genetics is when traits do not follow the Mendelian laws. Examples include the following:
  - incomplete dominance
  - codominance
  - polygenic inheritance
  - multiple alleles
  - linked genes

- Humans have 23 pairs of homologous chromosomes for a total of 26. Of these, 22 are autosomes, and two are designated as sex chromosomes.
  - Females have two X chromosomes; males have one X and one Y chromosome.
  - Because females have two X chromosomes, one is inactivated by condensation into Barr bodies.

# Chapter 10 Drill

Answers and explanations can be found in Chapter 16.

1. Two expecting parents wish to determine all possible blood types of their unborn child. If both parents have an AB blood type, which of the following blood types will their child NOT possess?

   (A) A
   (B) B
   (C) AB
   (D) O

2. A new species of tulip was recently discovered. A population of pure red tulips was crossed with a population of pure blue tulips. The resulting F1 generation was all purple. This result is an example of which of the following?

   (A) Complete dominance
   (B) Incomplete dominance
   (C) Codominance
   (D) Linkage

3. In pea plants, tall (T) is dominant over short (t) and green (G) is dominant over yellow (g). If a pea plant heterozygous for both traits is crossed with a plant that is recessive for both traits, approximately what percentage of the progeny plants will be tall and yellow?

   (A) 0%
   (B) 25%
   (C) 66%
   (D) 75%

Questions 4–6 refer to the following pedigree tree and paragraph.

Hemophilia is an X-linked disease associated with the inability to produce specific proteins in the blood-clotting pathway. Shown above is a family pedigree tree in which family members afflicted with the disease are shown with filled-in squares (male) or circles (females). A couple is trying to determine the likelihood of passing on the disease to their future children (represented by the ? symbol above) because the hemophilia runs in the woman's family.

4. Assuming that the woman is a carrier, what is the probability that the couple's first son will have hemophilia?

   (A) 0%
   (B) 25%
   (C) 50%
   (D) 100%

5. Why were both women in the family tree above free of the disease?

   (A) They were lucky because they didn't receive an X chromosome with the diseased allele.
   (B) They cannot get hemophilia because it is only associated with a diseased Y chromosome.
   (C) They are carriers and will only get the disease if they have children.
   (D) They have at least one normal (unaffected) copy of the X chromosome.

6. Turner syndrome is a disease in which an individual is born with only a single X chromosome. Suppose the woman in the couple is a carrier for hemophilia and has a child with Turner Syndrome, would this child have the disease?

 (A) Yes, because the child would only have one copy of the X chromosome and it would be affected.
 (B) No, because women cannot be affected by hemophilia.
 (C) No, because the child would have to receive a normal copy of the X chromosome from its mother.
 (D) Maybe, it depends on which X chromosome she receives from her mother.

7. Neuronal ceroid lipofuscinosis (NCL) is a group of autosomal recessive diseases characterized by blindness, loss of cognitive and motor function, and early death. One of the genes that are mutated in this disease is CLN3. When functional CLN3 protein is absent, neurons die because of increased storage material in the cells, presumably because the lysosomes aren't working properly. What is a valid explanation as to why parents of children with this devastating disease can be completely normal?

 (A) Parents of the autosomal recessive disease must both be carriers of the mutation on the CLN3 gene.
 (B) Carriers of the CLN3 mutation genotype do not show the phenotype because one normal CLN3 allele is present to provide a functioning CLN3 protein.
 (C) Redundant proteins take over the function of the mutant CLN3 protein in the parents.
 (D) X inactivation prevents expression of the CLN3 mutated protein in the parents.

8. A breeder of black Labrador puppies notices that in his line, a gene predisposing his dogs for white spots has arisen. He believes the white spot allele is autosomal recessive, and he wants to prevent it from continuing in his dogs. He sees that one of his female dogs, Speckle, has white spots, and he therefore assumes she is homozygous for the gene. But he must determine which males in his pack are heterozygous to avoid breeding them in the future. What is a reasonable plan?

 (A) Stop breeding Speckle.
 (B) Pair Speckle with a male with white spots.
 (C) Perform a testcross of Speckle with a black Labrador male. If any pups arise with white spots, stop breeding that male because he is heterozygous.
 (D) Breed Speckle with a chocolate Labrador.

9. Which of the following is a reason why certain traits do not follow Mendel's law of independent assortment?

 (A) The genes are linked on the same chromosome.
 (B) It only applies to eukaryotes.
 (C) Certain traits are not completely dominant.
 (D) Nondisjunction

10. The gene for coat color in some breeds of cats is found on the X chromosome. Calico cats are mottled with orange and black colorings. What is a possible explanation for the fact that true calico cats are only female?

 (A) The allele for coat color is randomly chosen by X-inactivation.
 (B) The coat color is linked with genes that are lethal to male cats, so male calico cats never live past birth.
 (C) Coat orange and black color is expressed only in female cats; males have only one color.
 (D) Male cats have Y chromosomes.

## REFLECT

Respond to the following questions:

- For what content topics discussed in this chapter do you feel you have achieved sufficient mastery to answer multiple-choice questions correctly?

- For which content topics discussed in this chapter do you feel you have achieved sufficient mastery to discuss effectively in a free response?

- For which content topics discussed in this chapter do you feel you need more work before you can answer multiple-choice questions correctly?

- For which content topics discussed in this chapter do you feel you need more work before you can discuss effectively in a free response?

- What parts of this chapter are you going to re-review?

- Will you seek further help, outside this book (such as a teacher, tutor, or AP Central), on any of the content in this chapter—and if so, on what content?

# Chapter 11
# Animal Structure
# and Function

# THE STRUCTURE AND FUNCTION OF ORGANISMS

To carry on with the business of life, higher organisms must all contend with the same basic challenges: obtaining nutrients, distributing them throughout their bodies, voiding wastes, responding to their environments, and reproducing. To accomplish these basic tasks, nature has come up with solutions. However, for all their differences, most animals have remarkably similar ways of dealing with these challenges.

While the AP Biology Exam does not include memorization questions about body systems, you'll likely see thematic questions which cover structure and function relationships, communication between cells and tissues, and interaction and interconnectedness of organismal biology. Although the AP Biology Course and Exam used to require memorization of each of the body systems, the curriculum (and, therefore, the exam) now requires only specific knowledge about the immune system, the nervous system, the endocrine system, and reproductive development because these systems best illustrate the communication and interaction between different cells within the organism as a whole. Questions discussing other systems will do so in the context of these themes. Therefore, this chapter covers these four key systems in detail, while giving you a solid foundation in the themes that you may see reflected in questions on the exam.

The systems we'll look at include the following:

- lymphatic/immune system
- nervous system
- endocrine system
- reproductive system, or morphogenesis

# THE LYMPHATIC AND IMMUNE SYSTEMS

In addition to the circulatory system, vertebrates have another system called the lymphatic system. The **lymphatic system** is made up of a network of vessels that conduct **lymph**, a clear, watery fluid formed from interstitial fluid. Lymph vessels are found throughout the body along the routes of blood vessels and play an important role in fluid homeostasis.

Thymus

Lymph node

Spleen

Lymph vessels

The lymphatic system has three functions.

- It collects, filters, and returns fluid to the blood by the contraction of adjacent muscles.

- It fights infection using lymphocytes, cells found in lymph nodes.

- It removes excess fluid from body tissue.

Sometimes a lymph vessel will form a **lymph node**, a mass of tissue found along the course of a lymph vessel. A lymph node contains a large number of lymphocytes:

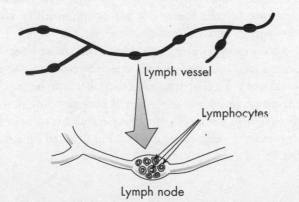

Lymph vessel

Lymphocytes

Lymph node

**Lymphocytes** are important in fighting infection. They multiply rapidly when they come in contact with an **antigen**, or foreign substance recognized by the **immune system**. (We'll talk about this in a second.) The lymph nodes swell when they're fighting an infection. That's why when you have a sore throat, one of the first things a doctor does is touch the sides of your throat to see if your lymph nodes are swollen, a probable sign of infection.

## The Immune System

The immune system, generally speaking, is one of the body's defense systems. It is a carefully and closely coordinated system of specialized cells, each of which plays a specific role in the war against bodily invaders. As we mentioned above, foreign molecules—be they viral, bacterial, or simply chemical—that can trigger an immune response are called antigens. The appearance of antigens in the body stimulates a defense mechanism that produces antibodies.

The body's first line of defense against foreign substances is the skin and mucous lining of the respiratory and digestive tracts. If these defenses are not sufficient, other nonspecific defense mechanisms are activated. These include **phagocytes** (which engulf antigens), **complement proteins** (which lyse the cell wall of the antigen), **interferons** (which inhibit viral replication and activate surrounding cells that have antiviral actions), and **inflammatory response** (a series of events in response to antigen invasion or physical injury).

## Types of Immune Cells

The primary cells of the immune system are lymphocytes: B-cells and T-cells. When an individual becomes infected by a **pathogen**, a disease-causing agent, B- and T- lymphocytes get activated. B-lymphocytes mature in bone marrow and are involved in the humoral response, which defends the body against pathogens present in extracellular fluids, such as lymphatic fluid or blood. When B-lymphocytes encounter pathogens, they are activated and produce clones. Some B-cells become memory cells that remain in circulation, allowing the body to mount a quicker response if a second exposure to the pathogen should occur. Other B-cells become plasma cells that produce antibodies, which are specific proteins that bind to the antigens on the surface of pathogens that originally activated them.

T-lymphocytes, maturing in the thymus, are involved in **cell-mediated immunity.** The plasma membrane of cells has **major histocompatibility complex (MHC) markers** that distinguish between self and foreign cells. When T-cells encounter cells infected by pathogens, they recognize the foreign antigen-MHC **markers** on the cell surface. Once activated, T-cells multiply and give rise to clones. Some T-cells become **memory T-cells,** whereas others become **helper T-cells.** Helper T-cells activate B-lymphocytes and other T-cells in responding to the infected cells. Memory T-cells recognize pathogens they have encountered before. Another type of T-cells, the **cytotoxic T-cell,** recognizes and kill infected cells.

## AIDS

**AIDS**, or "acquired immunodeficiency syndrome," is a devastating disease that interferes with the body's immune system. AIDS is caused by the specific infection of helper T-cells by human immunodeficiency virus (HIV). The virus essentially wipes out the helper T-cells, preventing the body from defending itself. Those afflicted with AIDS do not die of AIDS itself but rather of infections that they can no longer fight off due to their compromised immune systems.

One thing to remember about the immune cells and blood cells: All blood cells, white and red, are produced in the bone marrow. To summarize:

- *T-lymphocytes* actually fight infection and help the B-lymphocytes proliferate.

- *B-lymphocytes* produce *antibodies*.

# Lymphatic and Immune Systems Drill

Answers and explanations can be found in Chapter 16.

1. Lymph nodes are often evaluated when a patient is believed to have an active infection or cancer. Why?

    (A) Active infections or cancer often generate large quantities of fluids as cells rupture, the lymph nodes often swell as a result of the buildup of these fluids.

    (B) During an active infection, lymphocytes expand rapidly in lymph nodes causing swelling of the structures.

    (C) Lymph nodes are the most common tissue site of active infections or cancer.

    (D) Cancer often metastasizes to lymph nodes causing them to expand as new cancerous cells multiply.

Questions 2–4 refer to the following chart and paragraph.

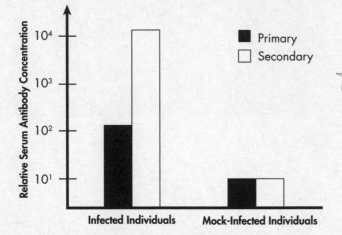

A scientist wishes to test the efficacy and immunogenicity of a newly developed live-attenuated virus vaccine strain. To test for immunogenicity, the scientist acquires 100 healthy volunteers and challenges them with either the vaccine strain or a placebo strain (consisting of isotonic saline solution). 12 days following the initial primary challenge, the serum antibody concentration is determined by an ELISA analysis. A follow-up challenge to test for existing memory was performed on day 60 using the same 100 individuals. Serum antibody titers were determined for the secondary challenge on day 72. These data are shown in the chart above.

2. The placebo strain was included in the experimental design in order to

    (A) evaluate the immune response to isotonic saline solution.

    (B) determine primary and secondary immune responses to the virus.

    (C) compare antibody titers of infected and mock-infected individuals as a control.

    (D) practice before challenging the same individuals with the vaccine strain.

3. What type of immune cells is responsible for generating the antibodies generated in this experiment?

    (A) Erythrocytes
    (B) B-Cells
    (C) T-Cells
    (D) Phagocytes

4. How can the scientist increase the statistical significance of these data?

    (A) They could repeat the experiment with a greater sample size.

    (B) They could use more control variables.

    (C) They could perform a third challenge on the same 100 individuals.

    (D) They could use a different statistical test, which provides the necessary statistical significance.

5. Myasthenia gravis is an autoimmune disorder in which patients experience fatigue, sluggishness, and weakness. Rogue autoantibodies in the bloodstream target the body's own acetylcholine receptors at the neuromuscular junctions, causing the patients' muscles to exhibit delayed responses. Which treatment would have the most immediate positive effect for patients of this disorder?

   (A) Broad-spectrum immunosuppression drugs to reduce the effectiveness of the immune system
   (B) Removal of the thymus because it is responsible for maturation of macrophages
   (C) Blood transfusion to remove or dilute the acetylcholine autoantibodies
   (D) Additional acetylcholine

6. Drugs that use the immune system to attack cancer cells have shown great promise in treating specific cancers like Hodgkin's lymphoma. With this type of drug, the doctors hope to permit the patient's own immune system to attack the abnormal cancer cells, resulting in shrinking tumors and remission in patients. Which type of cell are they likely mobilizing against the cancer?

   (A) Memory B-cells
   (B) Lymphocytes
   (C) Phagocytes
   (D) Natural T-killer cells

7. Innate immune system defenses in animals include skin, mucous membranes, and defensive enzymes present in saliva, tears, stomach acid, and skin. Analogous structures in plants could include

   (A) long stems
   (B) thorns and spines
   (C) attractive scents
   (D) stomata

8. When lymphocytes first encounter a pathogen, receptors on their surface first bind to the outside of the pathogen. Next, this activated lymphocyte multiplies and becomes numerous, allowing it to bind to many more pathogens of the same type that exist in the bloodstream. What are other results from this initial recognition of the pathogen?

   (A) Antibodies specific to the pathogen are secreted by B-cells.
   (B) The pathogen is destroyed by helper T-cells.
   (C) Only plasma B-cells remain.
   (D) The pathogen is permitted to infect more cells prior to its destruction.

9. After a macrophage engulfs a pathogen through phagocytosis, it destroys the pathogen and then the macrophage matures to become

   (A) T-cells that secrete specific antibodies for the pathogen
   (B) MHC antigen-presenting cells
   (C) activated plasma cells
   (D) memory B-cells

## THE NERVOUS SYSTEM

All organisms must be able to react to changes in their environment. As a result, organisms have evolved systems that pick up and process information from the outside world. The task of coordinating this information falls to the nervous system. The simplest nervous system is found in the hydra. It has a **nerve net** made up of a network of nerve cells, the impulse of which travels in both directions. As animals became more complex, they developed clumps of nerve cells called **ganglia**. These cells are like primitive brains. More complex organisms have a brain with specialized cells called **neurons**.

## Neurons

The functional unit in the nervous system is a neuron. That's because neurons receive and send the neural impulses that trigger organisms' responses to their environments. Let's talk about the parts of a neuron. A neuron consists of a **cell body**, **dendrites**, and an **axon**.

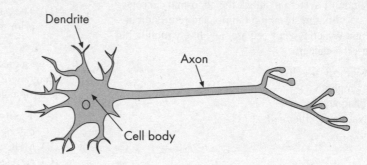

The cell body contains the nucleus and all the usual organelles found in the cytoplasm. Dendrites are short extensions of the cell body that receive stimuli. The axon is a long, slender extension that transmits an impulse from the cell body to another neuron or to an organ. A nerve impulse begins at the top of the dendrites, passes through the dendrites to the cell body, and moves down the axon.

## Types of Neurons

Neurons can be classified into three groups: **sensory neurons**, **motor (effector) neurons**, and **interneurons**. Sensory neurons receive impulses from the environment and bring them to the body. For example, sensory neurons in your hand are stimulated by touch. A motor neuron transmits the impulse to muscles or glands to produce a response. The muscle will respond by contracting, or the gland will respond by secreting a substance (such as a hormone). Interneurons are the links between sensory neurons and motor neurons. They're found in the brain or spinal cord:

## How Neurons Communicate

Before we talk about the events related to the transmission of a nerve impulse, let's review how neurons interact. There are billions of neurons running throughout the body, firing all the time. More often than not, one or more neurons are somewhat "connected." This means that one neuron has its dendrites next to another neuron's axon. In this way, the dendrites of one cell can pick up the impulse sent from the axon of another cell. The second neuron can then send the impulse to its cell body and down its axon, passing it on to yet another cell.

## Resting Potential

Neurons are not always transmitting signals. The transmission of an impulse depends on the ionic gradients that exist across the axonal membrane. In humans, the concentration of sodium ions is higher in the extracellular fluid than the concentration of potassium ions. The reverse is true inside the axonal membrane. The concentration of potassium ions is higher in the cytosol than the concentration of sodium ion. Because there are many potassium channels that are open but only a relatively small number of sodium channels open the resting potential results from the diffusion of sodium and potassium ions through open ions channels.

The resting potential arises from two activities:

- The $Na^+K^+$-ATPase pump—This pump pushes two potassium ions ($K^+$) into the cell for every three sodium ions ($Na^+$) it pumps out of the cell which leads to a net loss of positive charges within the cell.

- Leaky $K^+$ channels—Some potassium channels in the plasma membrane are "leaky" allowing a slow diffusion of $K^+$ out of the cell.

Both the $Na^+K^+$-ATPase pump and the leaky channels cause a potential difference between the inside of the neuron and the surrounding interstitial fluid. The resting membrane potential is always negative inside the cell, and the neuronal membrane is said to be **polarized**. In humans, the negative charge is –70mV.

Diffusion out

3Na⁺

Diffusion in

Na⁺/K⁺ pump

2K⁺

Neuron cytoplasm

## Action Potential

Here's what a neuron does in response to a stimulus. If a stimulus has enough intensity to excite a neuron, the cell reaches its **threshold**—the minimum amount of stimulus a neuron needs to respond. This creates what we call an **action potential**, that is, a change in the membrane potential that produces a nerve impulse. The action potential is an **all-or-none response**—it doesn't fire "part way."

## Depolarization

At the point where the axon connects to the cell body, tiny gated sodium ion channels open up and allow sodium ions to rush into the cell. So many sodium ions rush in that the cell now becomes more positive inside than outside. This change is known as **depolarization**: The interior of the cell has "switched" its polarity from a negative to a positive charge. For now, just remember that an action potential makes the cell depolarize.

The net change is substantial. The charge has shifted from the –70 mV we saw earlier to about +35 mV:

Let's recap what's happened so far. In an action potential:

- the cell's tiny "gates" open up
- sodium ions rush in
- the polarity of the cell changes; the axon is now positive on the inside and negative on the outside

## Repolarization

Once sodium ions have flooded the neuron, the sodium channels close. At this point, the potassium channels open. The potassium ions, which are on the inside of the axon, now rush out. As the potassium ions move out of the cell, the electrical charges reverse again. The inside of the cell becomes more negative than the outside of the cell.

We can now say that this section of the neuron has been **repolarized**. In other words, the charge has returned to its original polarization.

## The Refractory Period

Here's one thing you should remember: Although the charge has returned to its original state, at the end of the action potential the ions are now on the wrong side of the axonal membrane. Sodium ions are on the inside and potassium ions are on the outside of the axonal membrane. Originally, sodium ions were on the outside and potassium ions were on the inside. The neuron reestablishes the order of the ions, and this process is carried out by the **sodium-potassium pump**.

This pump reestablishes the original ion distribution by kicking three sodium ions out of the cell for every two potassium ions it brings into the cell. The period after an action potential is known as the **refractory period**.

During this period, the sodium channels are now reset and are able to open, but the cell membrane potential is further from the threshold. A greater stimulus is required to reach the threshold, so it is more difficult to initiate another action potential.

To summarize:

- A "resting" neuron is polarized; that is, it is more negative on the inside than on the outside.

- When an action potential comes along, the neuron transmits an impulse down its axon.

- First, voltage-gated sodium channels open, allowing sodium ions to rush in. This is known as *depolarization*: The neuron becomes more positive on the inside and more negative on the outside.

- Sodium channels close and potassium channels open, which restores its negative charge. This is known as *repolarization*.

- The neuron enters a refractory period.

- The neuron reestablishes the ion distribution thanks to the sodium-potassium pump.

When one small area is depolarized, it causes a "domino effect." The action potential spreads to the rest of the axon. The impulse is transmitted down the axonal membrane until it reaches the end of the axon called the **axon bulb**. Now the neuron wants to pass the impulse to the next neuron. How does it manage this?

When an impulse reaches the end of an axon, the axon releases a chemical called a **neurotransmitter** into the space between the two neurons. This space is called a **synapse**. The neurotransmitter diffuses across the synaptic cleft and binds to receptors on the dendrites of the next neuron:

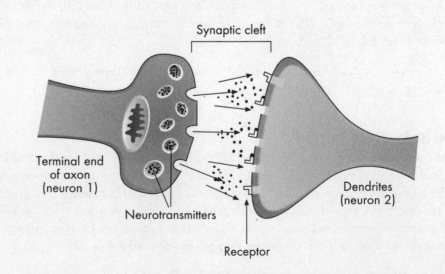

This usually triggers an action potential in the second neuron if the synaptic membrane is excited. Now the impulse moves along the second neuron from dendrites to axon:

Terminal end
of axon
(neuron 1)

Impulse

Dendrites
(neuron 2)

There are many neurotransmitters, but the most important one for the AP Biology Exam is called **acetylcholine**. Acetylcholine is a neurotransmitter that

- is released from the end of an axon, when $Ca^{2+}$ moves into the terminal end of the axon

- is picked up almost instantly by the dendrites of the next neuron

- can stimulate muscles to contract or inhibit postsynaptic potential

- is released between neurons in the parasympathetic system, which we will discuss shortly

The extra acetylcholine in the synaptic cleft is broken down by the enzyme **acetylcholinesterase**.

Other important neurotransmitters include **norepinephrine** and **GABA**. Norepinephrine is a peptide neurotransmitter that is released between neurons within the central nervous system. GABA is secreted in the central nervous system and acts as an inhibitor.

## Speed of an Impulse

Sometimes a neuron has supporting cells that wrap around its axon. These cells are called **Schwann cells**. Schwann cells produce a substance called the **myelin sheath**, which insulates the axon:

As you can see from the illustration above, the whole axon isn't covered with myelin sheaths. The spaces between myelin sheaths—the exposed regions of the axon—are called the **nodes of Ranvier**. Myelin sheaths speed up the propagation of an impulse. Instead of the standard "domino effect" that occurs during an action potential, the impulse can now jump from node to node. This form of conduction is called **saltatory conduction**.

Thanks to the myelin sheath, the neuron can transmit an impulse down the axon far more rapidly than it could without its help.

# Parts of the Nervous System

## Central Nervous System

The nervous system can be divided into two parts: the central nervous system and the peripheral nervous system.

All of the neurons within the brain and spinal cord make up the central nervous system. All of the other neurons lying outside the brain and the spinal cord—in our skin, our organs, and our blood vessels—are collectively part of the peripheral nervous system. Although both of these systems are really part of one system, we still use the terms *central* and *peripheral*.

So keep the following in mind:

- The **central nervous system** includes the neurons in the brain and spinal cord.

- The **peripheral nervous system** includes all the rest.

# Peripheral Nervous System

The peripheral nervous system is further broken down into the somatic nervous system and the autonomic nervous system.

- The **somatic nervous system** is the part that controls voluntary activities. For example, the movement of your eyes across the page as you read this line is under the control of your somatic nervous system.

- The **autonomic nervous system** is the part that controls involuntary activities. Your heartbeat and your digestive system, for example, are under the control of the autonomic nervous system.

The interesting thing about these two systems is that they sometimes overlap. For instance, you can control your breathing if you choose to. Yet most of the time you do not think about it: Your somatic system hands control of your respiration over to the autonomic system.

The autonomic system is broken down even further to the **sympathetic nervous system** and the **parasympathetic nervous system**. These two systems actually work antagonistically.

The sympathetic system controls the **"fight-or-flight" response**, which occurs when an organism confronted with a threatening situation prepares to fight or flee. To get ready for a quick, effective action, whether that be brawling or bolting, the sympathetic nervous system raises your heart and respiration rates, causes your blood vessels to constrict, increases the levels of glucose in your blood, and produces "goose bumps" on the back of your neck. It even reroutes your blood sugar to your skeletal muscles, in case you need to make a break for it. After the threat has passed, the parasympathetic nervous system brings the body back to **homeostasis**—that is, back to normal. It lowers your heart and respiratory rates and decreases glucose levels in the blood.

The flow chart below will give you a nice overview of the different parts of the nervous system:

# Parts of the Brain

The brain can also be divided into parts. Here's a summary of the major divisions within the brain.

| DIVISIONS WITHIN THE BRAIN | |
|---|---|
| Parts of the Brain | Function |
| **Cerebrum** | Controls all voluntary activities; receives and interprets sensory information; largest part of human brain |
| **Cerebellum** | Coordinates muscle activity and refinement of movement |
| **Hypothalamus** | Regulates homeostasis and secretes hormones; regulates pituitary gland |
| **Medulla** | Controls involuntary actions such as breathing, swallowing, heartbeat, and respiration |
| **Pons** | Connects parts of the brain with one another and contains respiratory center |
| **Midbrain** | Center for visual and auditory reflexes (pupil reflex and blinking) |
| **Thalamus** | Main sensory relay center for conducting information between the spinal cord and cerebrum |

The cerebrum consists of outer gray matter (the **cerebral cortex**) and inner white matter. One structure that is often mentioned on the AP Biology Exam is the **corpus callosum**. The corpus callosum is a thick band of nerve fibers of the white matter that enable the right and left side of the cerebral hemispheres to communicate.

# Nervous System Drill

Answers and explanations can be found in Chapter 16.

1. The image below depicts an action potential occurring in an active neuron. Which of the following best explains why the membrane potential depolarizes at –50 mV?

(A) At –50 mV, voltage-gated potassium channels open, allowing potassium ions (K⁺) to rush into the cell.

(B) At –50 mV, voltage-gated sodium channels open, allowing sodium ions (Na⁺) to rush into the cell.

(C) At –50 mV, voltage-gated potassium channels open, allowing potassium ions (K⁺) to rush out of the cell.

(D) At –50 mV, voltage-gated sodium channels open, allowing sodium ions (Na⁺) to rush out of the cell.

2. Fugu is a Japanese delicacy made from the preparation of pufferfish. Pufferfish produce a powerful toxin called tetrodotoxin, which binds to and blocks voltage-gated sodium channels. Specially trained Japanese chefs must carefully remove the toxin during preparation of fugu to prevent inadvertent poisoning by ingestion. Which of the following would likely occur to a neuron exposed to tetrodotoxin?

(A) The neuron would be able to send action potentials but would not be able to release neurotransmitters to neighboring neurons.

(B) The neuron would be able to send action potentials but would not be able to receive neurotransmitters from a presynaptic neuron.

(C) The neuron would not be able to send action potentials and would be unable to continue a neural response.

(D) The neuron would not be able to send action potentials but will still be depolarized by the influx of sodium ions.

3. Jane has recently stumbled upon two bear cubs while hiking a trail in the forest. Upon their discovery, the mother of the cubs came running out of the woods and charged toward Jane. Which of the following best describes the neural response in Jane?

(A) The sympathetic part of her autonomic nervous system was activated.

(B) The sympathetic part of her somatic nervous system was activated.

(C) The parasympathetic part of her autonomic nervous system was activated.

(D) The parasympathetic part of her somatic nervous system was activated.

4. An ER patient has recently been in a car accident and is having major difficulty coordinating walking and motion. However, all of their muscles appear to be functionally active. Which of the following parts of the brain was likely injured by the accident?

(A) Cerebrum

(B) Cerebellum

(C) Pons

(D) Hypothalamus

5. Ouabain is historically used by Somali tribes on poisoned arrows. Ouabain inhibits the activity of the sodium-potassium pump and is therefore used as a powerful neurotoxin. What effect would ouabain have on motor and sensory neurons in a victim of a poisoned arrow?

(A) The muscles would spasm because of an excess amount of calcium released from the sarcoplasmic reticulum.
(B) Action potentials would not be capable of firing because the pumps are blocked.
(C) Brain damage would result as the poison spread up the neurons to the central nervous system.
(D) Action potentials would fire in the affected neurons, and then repolarization would be impossible without functioning sodium-potassium pumps. Paralysis would result.

6. In a neuron, the majority of the protein manufacturing machinery will be found in the

(A) dendrites
(B) cell body
(C) axon
(D) axon terminal

7. Acetylcholinesterase is an enzyme hydrolase that destroys the neurotransmitter acetylcholine in the neuromuscular junction. Patients with defective acetylcholinesterase will have

(A) more easily relaxed musculature
(B) extreme weakness
(C) overactive and hyper-contraction of muscles
(D) no effect

8. In an action potential, the "threshold" is reached when

(A) enough voltage-gated channels open to produce a change in membrane polarity
(B) the synaptic vesicles release neurotransmitters into the synaptic cleft
(C) the peak depolarization occurs
(D) sodium ions enter the cell

9. Glutamine is the major stimulating neurotransmitter in the human brain. Neurons that oppose the effect of glutamine secrete which of the following to produce an inhibitory response?

(A) Epinephrine
(B) Nitric oxide
(C) Acetylcholine
(D) GABA

10. Oligodendricytes exist in the central nervous system, acting analogously to the Schwann cells of the peripheral nervous system. Therefore, their main objective is to

(A) act as immune cells of the brain
(B) produce a myelin sheath that insulates the axonal transmission
(C) direct depolarization of the membrane and perpetuate the nerve impulse
(D) transmit nerve impulses

11. Batten's disease is a childhood neurodegenerative disorder that causes blindness, cognitive decline, and ultimately premature death by the third decade of life. Patients exhibit atrophy of the cerebellum, as viewed by MRI scans. Which symptom would reflect cerebellum dysfunction or loss of a healthy cerebellum?

(A) Loss of voluntary activities; trouble initiating movements
(B) Difficulty with language and speech
(C) Loss of motor coordination
(D) Altered visual and auditory reflexes

# THE ENDOCRINE SYSTEM

Chemical messengers can be produced in one region of the body to act on target cells in another region. These chemicals, known as **hormones**, are produced in specialized organs called **endocrine glands**. Hormones have a number of functions including regulating growth, behavior, development, and reproduction. For example, the hormone **ecdysone** promotes molting and the metamorphosis of a larva to a butterfly. Ecdysone is stimulated to release when targeted by another insect hormone, **brain hormone**, in the prothoracic glands. Another hormone, **juvenile hormone**, causes larvae to retain their characteristics. Other chemical messengers are used for communication. For example, **pheromones** help animals to communicate with members of their species and attract the opposite sex.

An endocrine gland releases hormones directly into the bloodstream, which carries them throughout the body. Take a look at the endocrine glands in the human body:

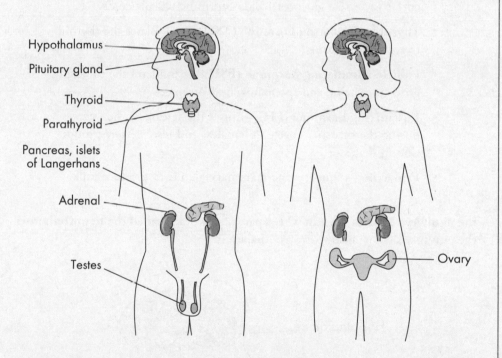

Before we launch into a review of the different hormones in the body, let's talk about how hormones work. Although hormones flow in your blood, they affect only specific cells. The cells that a hormone affects are known as the **target cells**. Suppose, for example, that gland X makes hormone Y. Hormone Y, in turn, has some effect on organ Z. We would then say that organ Z is the *target organ* of hormone Y.

Hormones also operate by a **negative feedback system**. That is, an excess of the hormone will signal the endocrine gland to temporarily shut down production. For example, when hormone Y reaches a peak level in the bloodstream, the organ secreting the hormone, gland X, will get a signal to stop producing hormone Y. Once the levels of hormone Y decline, the gland can resume production of the hormone.

# The Pituitary Gland

The **pituitary** is called the master gland because it releases many hormones that reach other glands and stimulate them to secrete their own hormones. The anterior pituitary therefore has many *target* organs. The pituitary has two parts: the **anterior pituitary** and the **posterior pituitary**. Each part secretes its own set of hormones.

The anterior pituitary secretes six hormones, three of which regulate growth and other organs. The other three are involved in regulating the reproductive systems. The hormones of the pituitary are:

- **Growth hormone (GH)**—stimulates growth throughout the body, targets bones and muscles

- **Adrenocorticotropic hormone (ACTH)**—stimulates the adrenal cortex to secrete glucocorticoids and mineralocorticoids

- **Thyroid-stimulating hormone (TSH)**—stimulates the thyroid to secrete thyroxine

- **Follicle-stimulating hormone (FSH)**—stimulates the follicle to grow in females, and spermatogenesis in males

- **Luteinizing hormone (LH)**—causes the release of the ovum during the menstrual cycle in females, and testosterone production in males

- **Prolactin**—stimulates the mammary glands to produce milk

The pituitary works in tandem with a part of the brain called the **hypothalamus**. The pituitary sits just below the hypothalamus:

Hypothalamus

Pituitary gland

The hypothalamus regulates the anterior pituitary by secreting neurohormones that can stimulate or inhibit the actions of the anterior pituitary. The other part of the pituitary, the posterior pituitary, secretes two hormones:

- **antidiuretic hormone (or vasopressin)**—regulates water intake by nephrons

- **oxytocin**—stimulates contraction of uterus and ducts of mammary glands

These hormones are actually made in the hypothalamus but are *stored* in the posterior pituitary. How about a mnemonic? When you think of the **pit**uitary, think of the GATOR pit:

**G**rowth hormone

**ACTH**

**T**hyroid-stimulating hormone

**O**xytocin

**R**, for antidiu**R**etic hormone (or vasopressin)

(All of which come from the **pit**uitary gland.)

And, to make things easier, the first three come first, or are "anterior" to the last two, which are "posterior." For us, that means that G, A, and T come from the anterior pituitary, while O and R come from the posterior pituitary.

As for the other three hormones, think of "FLAP." FSH, LH, and prolactin are all hormones that have to do with the reproductive system. Together, these two mnemonics should help you keep the different hormones of the pituitary gland straight. Now let's move on to the target organs.

## The Pancreas

We already know that the pancreas produces enzymes that it releases into the small intestine via the pancreatic duct. The pancreas also secretes two hormones, **glucagon (alpha cell)** and **insulin (beta cell)**, both of which are produced in clusters of cells called the **islets of Langerhans**. The target organs for these hormones are the liver and muscle cells. Glucagon, produced by $\alpha$ cells, stimulates the liver to convert **glycogen** into glucose and to release that glucose into the blood. Glucagon therefore *increases* the levels of glucose in the blood. Insulin has precisely the opposite effect that glucagon does.

When the blood has too much glucose floating around, insulin, produced by $\beta$ cells, allows body cells to remove glucose from the blood. Consequently, insulin *decreases* the level of glucose in the blood. Insulin is particularly effective on muscle and liver cells. In short:

- *Insulin* lowers the blood sugar level.
- *Glucagon* raises the blood sugar level.

## The Adrenal Glands

The adrenal glands contain two separate endocrine glands. One is called the **adrenal cortex**, and the other is called the **adrenal medulla**. Although they are part of the same organ, these two endocrine glands have very different effects on the body. Let's start by discussing the adrenal cortex.

## The Adrenal Cortex

Earlier we mentioned that adrenocorticotropic hormone, or simply ACTH, targets the adrenal cortex. When ACTH is released from the pituitary, it stimulates the adrenal cortex to produce and secrete its different hormones.

One group of hormones released by the adrenal cortex is the **glucocorticoids**. They increase the blood's concentration of glucose and help the body adapt to stress. In fact, glucocorticoids accomplish the same thing as glucagon, but they do so in a slightly different way. Glucocorticoids promote the conversion of amino acids and fatty acids to glucose.

Another set of hormones is the **mineralocorticoids**. They help the body retain $Na^+$ and water in the kidneys. They accomplish this by promoting the reabsorption of sodium ($Na^+$) and chlorine ($Cl^-$), which get together to form common salt (NaCl). When salt is retained, water soon follows.

So then, just remember that the adrenal cortex releases two types of hormones.

- Glucocorticoids target the liver and promote the release of glucose.
- Mineralocorticoids target the kidney and promote the retention of water.

## The Adrenal Medulla

The other adrenal gland, the adrenal medulla, is often referred to as the "emergency gland." It secretes two hormones: **epinephrine** and norepinephrine. These are the hormones involved in the fight-or-flight response. Both epinephrine and norepinephrine "kick in" under extreme stress. They increase your heart rate, metabolic rate, blood pressure, and give you a quick boost of energy.

## The Thyroid

The **thyroid gland**, which is located in the neck, is the target organ of the thyroid-stimulating hormone (TSH):

Thyroid

When the thyroid is stimulated by TSH, it releases the hormone **thyroxine**. Thyroxine, which contains iodine, is responsible for regulating the metabolic rate in your body tissues. Two conditions are associated with thyroid hormones. **Hyperthyroidism** occurs in individuals who regularly release too much thyroxine. They have a fast metabolic rate and tend to be irritable and nervous. On the other hand, individuals who suffer from **hypothyroidism** have too little thyroxine circulating in their bloodstream. They exhibit a slow metabolic rate and tend to be sluggish and overweight. To summarize:

> An individual with *hypo*thyroidism has a slow metabolic rate,
> while an individual with *hyper*thyroidism has a fast metabolic rate.

The thyroid also secretes another hormone called **calcitonin**. This hormone decreases your blood's concentration of calcium by concentrating free-floating calcium in the bones. You'll recall from our discussion of bone cells that bones contain collagen and calcium salts. Calcitonin is responsible for depositing these calcium molecules in the bones.

## The Parathyroids

The **parathyroids** are four little pea-shaped organs that rest on the thyroid. They secrete **parathyroid hormone**. Parathyroid hormone increases your blood calcium levels. Consequently, parathyroid hormone has the opposite effect that calcitonin does. If your blood needs more calcium, parathyroid hormone releases calcium ions stored in the bones. This process of building or breaking down bones to store and release calcium is called **bone remodeling**.

## Sex Hormones

Three hormones that we'll discuss in detail in the chapter on reproduction are **estrogen**, **progesterone**, and **testosterone**. Estrogen and progesterone are hormones released by the ovaries that regulate the menstrual cycle. Testosterone is the male hormone responsible for promoting spermatogenesis, the production of sperm. In addition, these hormones maintain secondary sex characteristics.

## How Hormones Work

How do hormones trigger the activities of their target cells? That all depends on whether the hormone is a steroid (lipid soluble) or a protein, peptide, or **amine** (not lipid soluble). If the hormone is a steroid, then the hormone can diffuse across the membrane of the target cell. It then binds to a receptor protein in the nucleus and activates specific genes contained in the DNA, which in turn make proteins.

However, if the hormone is a protein, peptide, or amine, it can't get into the target cell by means of simple diffusion. Remember: "Like dissolves like." The hormone binds to a receptor protein on the *cell membrane* of the target cell. In turn, this protein stimulates the production of a second messenger called **cyclic AMP (cAMP)**. The cAMP molecule then triggers various enzymes, leading to specific cellular changes.

Here's a summary of the hormones and their effects on the body.

| ORGAN | HORMONES | EFFECT |
|---|---|---|
| Anterior Pituitary | FSH | Stimulates activity in ovaries and testes |
| | LH | Stimulates activity in ovary (release of ovum) and production of testosterone |
| | ACTH | Stimulates the adrenal cortex |
| | Growth Hormone | Stimulates bone and muscle growth |
| | TSH | Stimulates the thyroid to secrete thyroxine |
| | Prolactin | Causes milk secretion |
| Posterior Pituitary | Oxytocin | Causes uterus to contract |
| | Vasopressin | Causes kidney to reabsorb water |
| Thyroid | Thyroid Hormone | Regulates metabolic rate |
| | Calcitonin | Lowers blood calcium levels |
| Parathyroid | Parathyroid Hormone | Increases blood calcium concentration |
| Adrenal Cortex | Aldosterone | Increases $Na^+$ and $H_2O$ reabsorption in kidneys |
| Adrenal Medulla | Epinephrine Norepinephrine | Increase blood glucose level and heart rate |
| Pancreas | Insulin | Decreases blood sugar concentration |
| | Glucagon | Increases blood sugar concentration |
| Ovaries | Estrogen | Promotes female secondary sex characteristics and thickens endometrial lining |
| | Progesterone | Maintains endometrial lining |
| Testes | Testosterone | Promotes male secondary sex characteristic and spermatogenesis |

While the nervous system and the endocrine system work in close coordination, there are significant differences between the two:

- The nervous system sends nerve impulses using neurons, whereas the endocrine system secretes hormones.

- Nerve impulses control rapidly changing activities, such as muscle contractions, whereas hormones deal with long-term adjustments.

# Endocrine System Drill

Answers and explanations can be found in Chapter 16.

1. Consider an experiment in which African green monkeys are injected with extract from the anterior pituitary gland. Which of the following would probably not be observed?

    (A) Increased production of ACH
    (B) Increased blood levels of glucocorticoids
    (C) Increased production of thyroxine
    (D) Increased production of vasopressin

2. Many hormones regulate functions by having antagonistic roles. Which of the following pairs of hormones result in opposite effects?

    (A) FSH and LH on testosterone production
    (B) Prolactin and oxytocin on milk production
    (C) Insulin and aldosterone on blood sugar concentrations
    (D) Calcitonin and parathyroid hormone (PTH) on blood calcium concentrations

3. All of the following hormones are produced by the thyroid or parathyroid hormones EXCEPT

    (A) thyroxine
    (B) thyroid stimulating hormone (TSH)
    (C) parathyroid hormone (PTH)
    (D) calcitonin

4. The hypothalamus is a critical structure for regulating hormonal levels. Which of the following is true of the hypothalamus?

    (A) It secretes thyroid-stimulating hormone (TSH).
    (B) It secretes luteinizing hormone (LH).
    (C) It is an extension of the pituitary gland.
    (D) It produces neurosecretory hormones.

5. Testosterone and estradiol are hormones that are derivatives of steroid molecules. Therefore, these hormones reach the interior of many different types of cells easily. Which of the following accurately describes this type of hormone reception?

    (A) Many types of cells have receptors for testosterone and estradiol on their plasma membrane.
    (B) These hormones are lipid-soluble, so they reach the interior of cells by crossing the plasma membrane.
    (C) Transport proteins on the surface of cells shuttle these hormones into the cell.
    (D) Hormones never enter cells, so this question must be inaccurate.

6. Which disorder is correctly matched with the dysfunctional cells?

    (A) Muscular dystrophy and bone cells
    (B) Giantism and the adrenal medulla
    (C) Diabetes and the pancreas
    (D) Dwarfism and the parathyroid gland

7. Analysis of a blood sample from a marathon runner at the end of a race will be expected to contain high levels of

    (A) glucagon
    (B) gastrin
    (C) insulin
    (D) cortisol

8. Only certain cells in the body are target cells for the hormone vasopressin. Which of the following best explains the specificity of this hormone?

(A) Target cells are the only cells exposed to vasopressin.
(B) Target cells are the only cells with receptors for vasopressin.
(C) Vasopressin enters all cells because it is nonpolar.
(D) Cells without vasopressin receptors destroy vasopressin before it can have an effect.

9. Hormones often exist as pairs with opposing effects. Which of the following accurately identifies opposing hormone pairs?

(A) Prolactin and oxytocin
(B) Calcitonin and thyroid hormone
(C) Estrogen and testosterone
(D) Insulin and glucagon

10. In type II diabetes, patients are no longer responsive to insulin, causing blood sugar levels to rise dangerously. What are potential causes for this loss of insulin response in adult patients?

(A) The pancreas does not produce enough insulin.
(B) Insulin receptors no longer bind with glucose.
(C) Insulin receptors are sequestered away from the surface of target cells, rendering them insensitive to insulin.
(D) The individual's glucagon levels are too high.

# THE REPRODUCTIVE SYSTEM AND EMBRYONIC DEVELOPMENT

Reproduction in animals involves the production of ova (egg cells) and sperm.

Before we launch into reproduction, let's take a look at something that ties in very nicely with everything we've just discussed about hormones and the endocrine system: the menstrual cycle.

Since the ovaries release hormones, they are considered endocrine glands. The ovaries have two main responsibilities:

- They manufacture **ova**.

- They secrete estrogen and progesterone, sex hormones that are found in females.

The hormones secreted by the ovaries are involved in the menstrual cycle.

## The Menstrual Cycle

### Phase 1: The Follicular Phase

In phase 1, the anterior pituitary secretes two hormones: **follicle-stimulating hormone (FSH)** and **luteinizing hormone (LH)**. The FSH stimulates several follicles in the ovaries to grow. Eventually, one of these follicles gains the lead and dominates the others, which soon stop growing. The one growing follicle now takes command.

Because the follicle is growing in this phase, the phase itself is known as the **follicular phase**. Remember that during all this time the follicle is releasing estrogen. Estrogen helps the uterine lining to thicken and eventually causes the pituitary to release LH. This increase in estrogen causes a sudden surge in luteinizing hormone. This release of LH is known as a **luteal surge**. LH triggers **ovulation**—the release of the follicle from the ovary.

There are thus three hormones associated with the follicular phase:

- Follicle-stimulating hormone (FSH)—originates in the pituitary gland.

- Estrogen—originates in the follicle.

- Luteinizing hormone (LH)—originates in the pituitary gland.

The luteal surge makes the follicle burst and release the ovum. The ovum then begins its journey into the **fallopian tube**, which is also known as the **oviduct**. This is a crucial event in the female menstrual cycle and is known as ovulation. Once the ovum has been released, the follicular phase ends and the ovum is ready to move on to the next phase.

In addition to the growth of the follicle, the follicular phase involves the thickening of the **uterine walls**, or **endometrium**. This happens in preparation for the implantation of a fertilized cell. The entire follicular phase lasts about 10 days.

## Phase 2: The Luteal Phase

By the end of the follicular phase, the ovum has moved into the fallopian tube and the follicle has been ruptured and left behind in the ovary. However, the ruptured follicle (now a fluid-filled sac) continues to function in the menstrual cycle. At this stage, it condenses into a little yellow blob called the **corpus luteum**, which is Latin for "yellow body."

The corpus luteum continues to secrete estrogen. In addition, it now starts producing the other major hormone involved in female reproduction, progesterone. Progesterone is responsible for readying the body for pregnancy. It does this by promoting the growth of glands and blood vessels in the endometrium. Without progesterone, a fertilized ovum cannot latch onto the uterus and develop into an embryo. We can therefore think of progesterone as the hormone of pregnancy.

After about 13 to 15 days, if fertilization and implantation have not occurred, the corpus luteum shuts down. Once it has stopped producing estrogen and progesterone, the final phase of the menstrual cycle begins.

## Phase 3: The Flow Phase, or Menstruation

Once the corpus luteum turns off, the uterus can no longer maintain its thickened walls. It starts to reabsorb most of the tissue that the progesterone encouraged it to grow. However, since there is too much to reabsorb, a certain amount is shed. This "sloughing off," or bleeding, is known as **menstruation**.

With the end of menstruation, the cycle starts all over again, readying the body for fertilization. Let's recap some of the major steps:

1. In the follicular phase, the pituitary releases FSH, causing the follicle to grow.

2. The follicle releases estrogen, which helps the endometrium to grow.

3. Estrogen causes the pituitary to release LH, resulting in a luteal surge.

4. This excess LH causes the follicle to burst, releasing the ovum during ovulation.

5. The shed follicle becomes the corpus luteum, which produces progesterone.

6. Progesterone, the "pregnancy hormone," enhances the endometrium, causing it to thicken with glands and blood vessels.

7. If fertilization does not occur after about two weeks, the corpus luteum dies, leading to menstruation—the sloughing off of uterine tissue.

If pregnancy occurs, the extraembryonic tissue of the fetus releases **human chorionic gonadotropin (HCG)**, which helps maintain the uterine lining. The AP Biology Exam is likely to contain questions dealing with the events that occur during the menstrual cycle and fertilization and where those events take place.

Take a look at the following diagram. Familiarize yourself with the parts of the female reproductive system and pay special attention to the different sites of the stages we've just discussed.

Remember:

- The follicles (and thus the ova) are contained in the ovaries.
- Hormones are released from the ovaries and pituitary gland.
- Fertilization occurs in the fallopian tube.
- The fertilized ovum implants itself in the uterus.

# The Male Reproductive System

Now let's discuss the hormones in the male reproductive system. Testosterone, along with the cortical sex hormones we saw earlier, is responsible for the development of the sex organs and secondary sex characteristics. In addition to the deepening of the voice, these characteristics include body hair, muscle growth, and facial hair, all of which indicate the onset of **puberty**. Testosterone also has another function. It stimulates the testes, the male reproductive organs, to manufacture **sperm cells**. Testosterone does this by causing cells in the testes to start undergoing meiosis.

Take a look at the male reproductive system.

Sperm and male hormones are produced in the testes. The main tissues of the testes, called the **seminiferous tubules**, are where spermatogonia undergo meiosis. The spermatids then mature in the **epididymis**. The **interstitial cells**, which are supporting tissue, produce testosterone and other androgens. Sperms then travel through the **vas deferens** and pick up fluids from the **seminal vesicles** (which provides them with fructose for energy) and the **prostate gland** (which provides an alkaline fluid that neutralizes the vagina's acidic fluids).

Unlike the female reproductive system, the male reproductive system continues to secrete hormones throughout the life of the male. FSH targets the seminiferous tubules of the testes, where it stimulates sperm production. LH stimulates interstitial cells to produce testosterone.

# Embryonic Development

How does a tiny, single-celled egg develop into a complex, multicellular organism? By dividing, of course. The cell will change shape and organization many times by going through a succession of stages. This process is called **morphogenesis**. In order for the human sperm to fertilize an egg, it must dissolve the corona radiata, a dense covering of follicle cells that surrounds the egg. Then the sperm must penetrate the zona pellucida, the zone below the corona radiata.

When an egg is fertilized by a sperm, it forms a diploid cell called a **zygote**:

Sperm        Ovum        Zygote

**Fertilization** triggers the zygote to go through a series of rapid cell divisions called **cleavage**. What's interesting at this stage is that the embryo doesn't grow. The cells just keep dividing to form a solid ball called a **morula**:

Zygote

Morula

One cell becomes two cells, two cells become four cells, and so on.

## Blastula

The next stage is called **blastula**. As the cells continue to divide, they "press" against each other and produce a fluid-filled cavity called a **blastocoel**:

Blastula (cross section)

## Gastrula

During **gastrulation**, the zygote begins to change its shape. Cells now migrate into the blastocoel and differentiate to form three germ layers: the ectoderm, mesoderm, and endoderm.

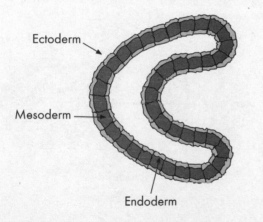

Gastrula (cross-section)

- The outer layer becomes the ectoderm.
- The middle layer becomes the mesoderm.
- The inner layer becomes the endoderm.

Each germ layer gives rise to various organs and systems in the body. Here's a list of the organs that develop from each germ layer.

- The ectoderm produces the epidermis (the skin), the eyes, and the nervous system.

- The endoderm produces the inner linings of the digestive tract and respiratory tract, as well as accessory organs such as the pancreas, gall bladder, and liver. These are called "accessory" organs because they are offshoots of the digestive tract, as opposed to the channels of the tract itself.

- The mesoderm gives rise to everything else. This includes bones and muscles as well as the excretory, circulatory, and reproductive systems.

## Organogenesis

The **neurula** stage begins with the formation of two structures: the **notochord**, a rod-shaped structure running beneath the nerve cord, and the **neural tube** cells, which develop into the central nervous system. By the end of this stage, we're well on our way to developing a nervous system:

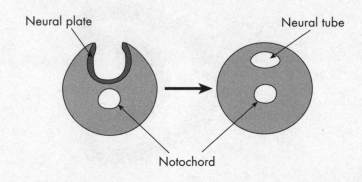

As far as the AP Biology Exam is concerned, the *order* of the stages and the various events is extremely important. For our purposes, think of the order of embryological development in this way:

Zygote → Cleavage → Blastula → Gastrula → Organogenesis

## What About Chicken Embryos?

In addition to the primary germ layers, some animals have **extraembryonic membranes**. Your average developing chicken, for example, possesses these membranes.

There are basically four extraembryonic membranes: the **yolk sac**, **amnion**, **chorion**, and **allantois**. These extra membranes are common in birds and reptiles.

For the AP Biology Exam, you should be familiar with these membranes and their functions.

| FUNCTIONS OF EXTRAEMBRYONIC MEMBRANES ||
|---|---|
| **Extraembryonic membrane** | **Function** |
| Yolk sac | Provides food for the embryo |
| Amnion | Forms a fluid-filled sac that protects the embryo |
| Allantois | Membrane involved in gas exchange; stores uric acid |
| Chorion | Outermost membrane that surrounds all the other extraembryonic membranes |

## Fetal Embryo

The fetal **embryo** also has extraembryonic membranes during development: the amnion, chorion, allantois, and yolk sac. The **placenta** and the **umbilical cord** are outgrowths of these membranes. The placenta is the organ that provides the **fetus** with nutrients and oxygen and gets rid of the fetus's wastes. The placenta develops from both the chorion and the uterine tissue of the mother. The umbilical cord is the organ that connects the embryo to the placenta.

## Development of an Embryo

During embryonic development, some tissues determine the fate of other tissues in a process called **induction**. Certain cells, called **organizers**, release a chemical substance (a **morphogen**) that moves from one tissue to the target tissue. It is now known that development involves many episodes of embryonic induction.

**Homeotic genes** control the development of the embryo. Some homeotic genes, called **homeobox genes**, consist of homeoboxes (short, nearly identical DNA sequences) that encode proteins that bind to DNA; these proteins tell cells in various segments of the developing embryo what type of structures to make. The process by which a less specialized cell becomes a more specialized cell type is called **differentiation**. Interestingly, homeobox genes are shared by almost all eukaryotic species. **Hox genes**, which are a subset of homeobox genes, specify the position of body parts in the developing embryo. Mutations in Hox genes result in the conversion of one body part into another. For example, in Drosophila, a specific Hox mutation results in a leg developing where an antenna would normally be.

The cytoplasm can also have an influence on embryonic development. For example, chicken and frog embryos contain more yolk in one pole (the vegetal pole) versus the other pole (the animal pole). This causes cells within the animal pole to divide more, and thus to be smaller than those of the vegetal pole. Because of the distribution of the yolk, cleavage of the egg does not produce eggs that develop normally. If the egg is divided into an animal and vegetal pole, development does not proceed normally.

**Apoptosis,** or programmed cell death, plays a crucial role in normal differentiation and development. For example, in a human embryo, apoptosis allows for the removal of tissue between newly developing fingers and toes to allow for separate digits. If this process is absent or incomplete, an individual is born with syndactyly, which is webbed hands or feet.

# Reproductive System and Embryonic Development Drill

Answers and explanations can be found in Chapter 16.

1. Shown below is the cross-section structure of a gastrula. Which of the following anatomical structures will ultimately be derived from the ectoderm?

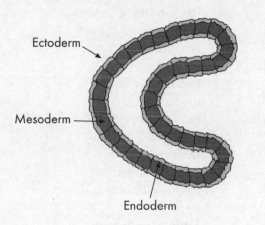

   Ectoderm

   Mesoderm

   Endoderm

   (A) GI tract
   (B) Brain
   (C) Heart
   (D) Pancreas

2. The menstrual cycle is critical for maintaining a state of readiness to prepare for impregnation and implantation of a developing embryo. Which of the following structures is essential for producing progesterone, which protects and enhances the endometrium?

   (A) Ova
   (B) Corpus luteum
   (C) Uterus
   (D) Pituitary gland

3. The male prostate gland performs an important role in producing a fluid, which is added to sperm before being released. What primary function does the prostate secretions perform?

   (A) Providing nutrients such as fructose for energy
   (B) Providing growth factors which are required for the soon-to-be developing embryo
   (C) Providing alkaline salts which neutralize the acidic conditions in the vagina and uterus
   (D) Providing a chemical signal to the testes to promote spermatogenesis

4. Embryos from many different organisms share several common structures. For instance, humans have a structure named a "gill slit" early during embryological development, homologous to a similar structure in bird and fish embryos. What is a possible explanation for the presence of this structure?

   (A) The sequence of embryological development is quite conserved across many different types of animals and species because we evolved from a common ancestor.
   (B) Embryos need the gill slits to breathe *in utero*.
   (C) Because we evolved from birds, we have similar development.
   (D) Gill slits are present to establish patterns of cells for our respiratory system to develop.

5. *C. elegans* is a roundworm genetic model with almost 1,000 cells. From the zygote, the cell lineage of each cell in the adult organism has been determined by fate mapping. While following the lineage of each cell in *C. elegans*, researchers found that over 100 cells die by apoptosis in normal development of the adult roundworm. Why do cells normally die by apoptosis in development?

   (A) Cleavage occurs early in embryonic development to produce too many cells.
   (B) Fingers and toes develop as individual digits by apoptosis.
   (C) Many neurons divided in the early developing nervous system, but only the neurons with strong connections to other neurons survive to later stages of development.
   D) *C. elegans* divides many cells during development. Excess cells are shed by the adult worm.

# KEY TERMS

## Lymphatic and Immune Systems

lymphatic system
lymph
lymph node
lymphocytes
antigen
immune system
phagocytes
complement proteins
interferons
inflammatory response
pathogen
cell-mediated immunity
major histocompatibility complex markers (MHC markers)
memory T-cells
helper T-cells
cytotoxic T-cell
AIDS
T-lymphocytes
B-lymphocytes

## Nervous System

nerve net
ganglia
neurons
cell body
dendrites
axon
sensory neurons
motor (effector) neurons
interneurons
polarized
threshold
action potential
all-or-none response
depolarization
repolarized
sodium-potassium pump
refractory period
axon bulb
neurotransmitter
synapse
acetylcholine
aceytlcholinesterase
norepinephrine
GABA

Schwann cells
myelin sheath
nodes of Ranvier
saltatory conduction
central nervous system
peripheral nervous system
somatic nervous system
autonomic nervous system
sympathetic nervous system
parasympathetic nervous system
"fight-or-flight" response
homeostasis
cerebrum
cerebellum
hypothalamus
medulla
pons
midbrain
thalamus
cerebral cortex
corpus callosum

## Endocrine System

hormones
endocrine glands
ecdysone
brain hormone
juvenile hormone
pheromones
target cells
negative feedback system
pituitary
anterior pituitary
posterior pituitary
growth hormone (GH)
adrenocorticotropic hormone (ACTH)
thyroid-stimulating hormone (TSH)
follicle-stimulating hormone (FSH)
luteinizing hormone (LH)
prolactin
hypothalamus
antidiuretic hormone (or vasopressin)
oxytocin
glucagon (alpha cell)
insulin (beta cell)
islets of Langerhans
glycogen

adrenal cortex
adrenal medulla
glucocorticoids
mineralocorticoids
epinephrine
thyroid gland
thyroxine
hyperthyroidism
hypothyroidism
calcitonin
parathyroids
parathyroid hormone
bone remodeling
estrogen
progesterone
testosterone
amine
cyclic AMP (cAMP)

**Reproductive System and Embryonic Development**
ova
follicle-stimulating hormone (FSH)
luteinizing hormone (LH)
follicular phase
luteal stage
ovulation
fallopian tube
oviduct
uterine walls
endometrium
corpus luteum
menstruation
human chorionic gonadotropin (HCG)
puberty

sperm cells
seminiferous tubules
epididymis
interstitial cells
vas deferens
seminal vesicles
prostate gland
morphogenesis
zygote
fertilization
cleavage
morula
blastula
blastocoel
gastrulation
neurula
notochord
neural tube
extraembryonic membranes
yolk sac
amnion
chorion
allantois
embryo
placenta
umbilical cord
fetus
induction
organizers
morphogen
homeotic genes
homebox genes
differentiation
Hox genes
apoptosis

# Summary

- The AP Biology Exam requires specific knowledge of only four systems: immune system, nervous system, endocrine system, and reproduction/development. Other body systems might appear on the test, but only in the context of the kinds of themes covered in this chapter: structure and function relationships, communication between cells, and so on. Understanding the following key points about the four systems highlighted in this chapter will help you with the thematic questions about various body systems on the test.

## Immune System

- The immune system is the defense system against pathogens that enter the organism. It is divided into two systems:
  - The innate immune system is the first response, the initial protection, and it includes the skin and mucous linings. Phagocytes and complement proteins also target this initial response to neutralize the invader.
  - Acquired immunity can be cell-mediated or an-tibody-based. T-lymphocytes include cytotoxic killer T-cells and helper T-cells. B-cells include plasma and memory B-cells.

## Nervous System

- Neurons are highly specialized, terminally differentiated cells. They are composed of dendrites, cell bodies, axons, and synaptic terminals.
  - Neurons are classified into three groups: sensory neurons, motor neurons, and interneurons.
  - Neuronal transmission is based on action potentials.
  - The junction between neurons is called a synapse. The pre-synaptic axon releases neurotransmitters to the post-synaptic dendrite to perpetuate the transmission.
  - Myelin speeds the transmission along axons.
- The central nervous system is composed of the brain and spinal cord.
- The peripheral nervous system senses and responds to stimuli, and it is composed of the somatic (voluntary) nervous system and the autonomic (involuntary) nervous system, which includes the sympathetic and parasympathetic nervous systems.

## Endocrine System

- Endocrine glands are tissues or organs that excrete hormones. Hormones mediate growth, reproduction, waste disposal, nutrient absorption, and behavior.

- The pituitary gland is the master gland.

- The hypothalamus regulates the anterior pituitary and makes hormones that are stored in the posterior pituitary. (Remember the GATOR pit!)

- Target cells receive the specific hormone via receptors either on the surface of the cell or internally, and react through signal transduction and response.

## Reproduction/Development

- Gametes from males (sperm) and females (ova) combine during fertilization to create a zygote.

- The zygote then goes through cleavage resulting to form a morula, blastula, gastrula, and ultimately an embryo.

- In chicken embryos there are additional extra embryonic membranes, the yolk sac, the amnion, the chorion, and the allantois.

- Development is controlled by highly conserved *Hox* genes, which provide directionality and polarity to the developing embryo.

- Apoptosis is also critical to development.

## REFLECT

Respond to the following questions:

- For what content topics discussed in this chapter do you feel you have achieved sufficient mastery to answer multiple-choice questions correctly?

- For which content topics discussed in this chapter do you feel you have achieved sufficient mastery to discuss effectively in a free response?

- For which content topics discussed in this chapter do you feel you need more work before you can answer multiple-choice questions correctly?

- For which content topics discussed in this chapter do you feel you need more work before you can discuss effectively in a free response?

- What parts of this chapter are you going to re-review?

- Will you seek further help, outside this book (such as a teacher, tutor, or AP Central), on any of the content in this chapter—and if so, on what content?

# Chapter 12
# Evolution

All of the organisms we see today arose from earlier organisms. This process, known as evolution, can be described as a change in a population over time. Interestingly, however, the driving force of **evolution**, **natural selection**, operates on the level of the individual. In other words, evolution is defined in terms of populations but occurs in terms of individuals.

## NATURAL SELECTION

What is the basis of our knowledge of evolution? Much of what we now know about evolution is based on the work of **Charles Darwin**. Darwin was a nineteenth-century British naturalist who sailed the world in a ship named the HMS *Beagle*. Darwin developed his theory of evolution based on natural selection after studying animals in the Galápagos Islands and other places.

Darwin concluded that it was impossible for the finches and tortoises of the Galápagos simply to "grow" longer beaks or necks. Rather, the driving force of evolution must have been natural selection. Quite simply put, this means that nature would "choose" which organisms survive on the basis of their fitness. For example, on the first island Darwin studied, there must once have been short-necked tortoises. Unable to reach the higher vegetation, these tortoises eventually died off, leaving only those tortoises with longer necks. Consequently, evolution has come to be thought of as "the survival of the fittest": Only those organisms most fit to survive will survive and reproduce.

Darwin elaborated his theory in a book entitled *On the Origin of Species*. In a nutshell, here's what Darwin observed:

- Each species produces more offspring than can survive.

- These offspring compete with one another for the limited resources available to them.

- Organisms in every population vary.

- The fittest offspring, or those with the most favorable traits, are the most likely to survive and therefore produce a second generation.

## Lamarck and the Long Necks

Darwin was not the first to propose a theory explaining the variety of life on Earth. One of the most widely accepted theories of evolution in Darwin's day was that proposed by **Jean-Baptiste de Lamarck**.

In the eighteenth century, Lamarck had proposed that *acquired* traits were inherited and passed on to offspring. For example, in the case of our giraffes, Lamarck's theory said that the giraffes had long necks because they were constantly reaching for higher leaves while feeding. This theory is referred to as the "law of use and disuse," or, as we might say now, "use it or lose it." According to Lamarck, giraffes have long necks because they constantly *use* them.

We know now that Lamarck's theory was wrong: Acquired changes—that is, changes at a "macro" level in somatic (body) cells—cannot be passed on to germ cells. For example, if you were to lose one of your fingers, your children would not inherit this trait.

## Evidence for Evolution

In essence, nature "selects" which living things survive and reproduce. Today, we find support for the theory of evolution in several areas:

- **Paleontology**, or the study of fossils. Paleontology has revealed to us both the great variety of organisms (most of which, including trilobites, dinosaurs, and the woolly mammoth, have died off) and the major lines of evolution.

- **Biogeography**, or the study of the distribution of **flora** (plants) and **fauna** (animals) in the environment. Scientists have found related species in widely separated regions of the world. For example, Darwin observed that animals in the Galápagos have traits similar to those of animals on the mainland of South America. One possible explanation for these similarities is a common ancestor. As we'll see below, there are other explanations for similar traits. However, when organisms share multiple traits, it's pretty safe to say that they also shared a common ancestor.

- **Embryology**, or the study of the development of an organism. If you look at the early stages in vertebrate development, all the embryos look alike! All vertebrates—including fish, amphibians, birds, and even humans—show fishlike features called gill slits.

- **Comparative anatomy**, or the study of the anatomy of various animals. Scientists have discovered that some animals have similar structures that serve different functions. For example, a human's arm, a dog's leg, a bird's wing, and a whale's fin are all the same appendages, though they have evolved to serve different purposes. These structures, called **homologous structures**, also point to a common ancestor.

| Human's arm | Dog's leg | Bird's wing | Whale's fin |

In contrast, sometimes animals have features with the same function but that are structurally different. A bat's wing and an insect's wing, for example, are both used to fly. They therefore have the same function, but they have evolved totally independently of one another. These are called **analogous structures**. Another classic example of an analogous structure is the eye. Though scallops, insects, and humans all have eyes, these three different types of eyes are thought to have evolved entirely independently of one another. They are therefore analogous structures.

- **Molecular biology**. Perhaps the most compelling proof of all is the similarity at the molecular level. Today, scientists can examine the nucleotide and amino acid sequences of different organisms. From these analyses, we've discovered that organisms that are closely related have a greater proportion of sequences in common than distantly related species. For example, most of us don't look much like chimpanzees. However, by some estimates, as much as 99% of our genetic code is identical to that of a chimp.

## GENETIC VARIABILITY

In Chapter 8, we saw how traits are passed from parents to offspring. You'll recall from our discussion of heredity that different alleles are passed from parents to their progeny. For example, you might have an allele for brown eyes from your mother and an allele for blue eyes from your father. Since brown is dominant, you'll wind up with brown eyes. We also saw how these alleles are in fact just different forms of the same gene.

As you know, no two individuals are identical. The differences in each person are known as **genetic variability**. All this means is that no two individuals in a population have identical sets of alleles (except, of course, identical twins). In fact, the survival of a species is dependent on this genetic variation; it allows a species to survive in a changing environment. How did all this wonderful variation come about? Through **random mutation**.

It might be hard to think of it in this way, but this is the very foundation of evolution, as we'll soon see. Now that we've reintroduced genes, we can refine our definition of evolution. More specifically:

> Evolution is the change in the gene pool of a population over time.

## The Peppered Moths

Let's take an example. During the 1850s in England, there was a large population of peppered moths. In most areas, exactly half of them were dark or carried alleles for dark coloring. The other half carried the alleles for light coloring. This 1:1 ratio of phenotypes was observed until air pollution, due primarily to the burning of coal, changed the environment. What happened?

Imagine two different cities: City 1 (in the south) and City 2 (in the north). Prior to the Industrial Revolution, both of these cities had unpolluted environments. In both of these environments, dark moths and light moths lived comfortably side-by-side. For simplicity's sake, let's say our proportions were a perfect fifty-fifty, half dark and half light. However, at the height of the Industrial Revolution, City 2, our northern city, was heavily polluted, whereas City 1, our southern city, was unchanged. In the north, where all the trees and buildings were thick with soot, the light moths didn't stand a chance. They were impossible for a predator to miss! As a result, the predators gobbled up light-colored moths just as fast as they could reproduce, sometimes even before they reached an age in which they *could* reproduce. However, the dark moths were just fine. With all the soot around, the predators couldn't even see them; they continued doing their thing—above all, *reproducing*. And when they reproduced, they had more and more offspring carrying the dark allele.

After a few generations, the peppered moth gene pool in City 2 changed. Although our original moth gene pool was 50 percent light and 50 percent dark, excessive predation changed the population's genetic makeup. By about 1950, the gene pool reached 90 percent dark and only 10 percent light. This occurred because the light moth didn't stand a chance in an environment where it was so easy to spot. The dark moths, on the other hand, multiplied just as fast as they could.

In the southern city, you'll remember, there was very little pollution. What happened there? Things remained pretty much the same. The gene pool was unchanged, and the population continued to have roughly equal proportions of light moths and dark moths.

# CAUSES OF EVOLUTION

Natural selection, the evolutionary mechanism that "selects" which members of a population are best suited to survive and which are not, works both "internally" and "externally": *internally* through random mutations and *externally* through environmental pressures.

To see how this process unfolds in nature, let's return to the moth case. Why did the dark moths in the north survive? Because they were dark-colored. But how did they become dark-colored? The answer is, through random mutation. One day, a moth was born with dark-colored wings. As long as a mutation does not kill an organism before it reproduces (most mutations, in fact, do), it may be passed on to the next generation. Over time, this one moth had offspring. These, too, were dark. The dark- and the light-colored moths lived happily side by side until something from the outside—in our example, the environment—changed all that.

The initial variation came about by chance. This variation gave the dark moths an edge. However, the edge did not become apparent until something made it apparent. In our case, that something was the intensive pollution due to the burning of coal. The abundance of soot made it easier for predators to spot the light-colored moths, thus effectively removing them from the population. Therefore, dark color is an **adaptation**, a variation favored by natural selection.

Eventually, over long stretches of time, these two different populations might change so much that they could no longer reproduce together. At that point, we would have two different species, and we could say, definitively, that the moths had evolved. As a consequence of random mutation and the pressure put on the population by an environmental change, evolution occurred.

## Types of Selection

The situation with our moths is an example of **directional selection**. One of the phenotypes was favored at one of the extremes of the normal distribution.

In other words, directional selection "weeds out" one of the phenotypes. In our case, dark moths were favored and light moths were practically eliminated. Here's one more thing to remember: Directional selection can happen only if the appropriate allele—the one that is favored under the new circumstances—is already present in the population. Two other types of selection are **stabilizing selection** and **disruptive selection**.

Stabilizing selection means that organisms in a population with extreme traits are eliminated. This type of selection favors organisms with common traits. It "weeds out" the phenotypes that are less adaptive to the environment. A good example is birth weight in human babies. An abnormally small baby has a higher chance of having birth defects; conversely an abnormally large baby will have a challenge in terms of a safe birth delivery.

Disruptive selection, on the other hand, does the reverse. It favors both the extremes and selects *against* common traits. For example, females are "selected" to be small and males are "selected" to be large in elephant seals. You'll rarely find a female or male of intermediate size.

**Types of Selection**

## SPECIES

A dog and a bumblebee obviously cannot come together to produce offspring. Therefore, they are different **species**. However, a poodle and a Great Dane could reproduce (at least in theory). We would not say that they are different species; they are merely different breeds.

Let's get back to our moths. We said earlier that evolution occurred when they could no longer reproduce. In fact, this is simply the endpoint of that particular cycle of evolution: **speciation**. Speciation refers to the emergence of new species. The type of evolution that our peppered moths underwent is known as **divergent evolution**.

Divergent evolution results in closely related species with different behaviors and traits. As with our example, these species often originate from a common ancestor. More often than not, the "engine" of evolution is cataclysmic environmental change, such as pollution in the case of the moths. Geographical barriers, new stresses, disease, and dwindling resources are all factors in the process of evolution. Pre- and post-zygotic barriers also prevent organisms of two different species from mating to produce viable offspring. **Pre-zygotic barriers** prevent fertilization. Examples of this kind of barrier include temporal isolation, or when two species reproduce at different times of the year. A **post-zygotic barrier** is related to the inability of the hybrid to produce offspring. For example, a horse and a donkey can mate to produce a mule, but mules are sterile and therefore cannot produce a second generation.

**Convergent evolution** is the process by which two unrelated and dissimilar species come to have similar (analogous) traits, often because they have been exposed to similar selective pressures. Examples of convergent evolution include aardvarks, anteaters, and pangolins. They all have strong, sharp claws and long snouts with sticky tongues to catch insects, yet they evolved from three completely different mammals.

There are two types of speciation: **allopatric speciation** and **sympatric speciation**. Allopatric speciation simply means that a population becomes separated from the rest of the species by a geographical barrier so that they can't interbreed. An example would be a mountain that separates two populations of ants. In time, the two populations might evolve into different species. If, however, new species form without any geographic barrier, it is called sympatric speciation. This type of speciation is common in plants. Two species of plants may evolve in the same area without any geographic barrier.

## POPULATION GENETICS

Mendel's laws can also extend to the population level. Suppose you caught a bunch of fruit flies—about 1,000. Let's say that 910 of them were red-eyed and 90 were green-eyed. If you allowed the fruit flies to mate and counted the next generation, we'd see that the ratio of red-eyed to green-eyed fruit flies would remain the same: 91 percent red-eyed and 9 percent green-eyed. That is, the allele frequency would remain constant. At first glance you may ask, how could that happen?

The **Hardy-Weinberg law** states that even with all the shuffling of genes that goes on, the relative frequencies of genotypes in a population still prevail over time. The alleles don't get lost in the shuffle. The dominant gene doesn't become more prevalent, and the recessive gene doesn't disappear.

Let's say that the allele for red eyes, R, is dominant over the allele for green eyes, r. Red-eyed fruit flies include homozygous dominants, RR, and heterozygous, Rr. The green-eyed fruit flies are recessive, rr.

## Hardy-Weinberg Equations

The frequency of each allele is described in the equation below. The allele must be either R or r. Let "$p$" represent the frequency of the R allele and "$q$" represent the frequency of the other allele in the population.

$$p + q = 1$$

This sum of the frequencies must add up to one. If you know the value of one of the alleles, then you'll also know the value of the other allele.

We can also determine the frequency of the *genotypes* in a population using the following equation:

$$p^2 + 2pq + q^2 = 1$$

In this equation, $p^2$ represents the homozygous dominants, $2pq$ represents the heterozygotes, and $q^2$ represents the homozygous recessives.

So how do we use these equations? Use the proportions in the population to figure out both the allele and genotype frequencies. Let's calculate the frequency of the genotype for green-eyed fruit flies. If 9 percent of the fruit flies are green-eyed, then the *genotype* frequency, $q^2$, is 0.09. You can now use this value to figure out the frequency of the recessive allele in the population. The allele frequency for green eyes is equal to the square root of 0.09—that's 0.3. If the recessive allele is 0.3, the dominant allele must be 0.7. That's because 0.3 + 0.7 equals 1.

Using the second equation, you can calculate the genotypes of the homozygous dominants and the heterozygotes. The frequency for the homozygous dominants, $p^2$, is 0.7 × 0.7, which equals 0.49. The frequency for the heterozygotes, $2pq$, is 2 × 0.3 × 0.7, which equals 0.42. If you include the frequency of the recessive genotype—0.09—the numbers once again add up to 1.

## Hardy-Weinberg Equilibrium

The **Hardy-Weinberg law** says that a population will be in genetic equilibrium only if it meets these five conditions: 1) a large population, 2) no mutations, 3) no immigration or emigration, 4) random mating, and 5) no natural selection.

## Violations of the Hardy-Weinberg Law

When these five conditions are met, the gene pool in a population is pretty stable. Any departure from them results in changes in allele frequencies in a population. For example, if a small group of your fruit flies moved to a new location, the allele frequency may be altered and result in evolutionary changes. That's an example of **genetic drift** called the founder effect. In other words, the gene frequency may differ from the original gene pool. Genetic drift often occurs in new colonies with small populations.

# KEY TERMS

evolution
natural selection
Charles Darwin
Jean-Baptiste de Lamarck
paleontology
biogeography
flora
fauna
embryology
comparative anatomy
homologous structures
analogous structures
molecular biology
genetic variability
random mutation

adaptation
directional selection
stabilizing selection
disruptive selection
species
speciation
divergent evolution
pre-zygotic barriers
post-zygotic barriers
convergent evolution
allopatric speciation
sympatric speciation
Hardy-Weinberg law
genetic drift

# Summary

- Charles Darwin made the following key observations that led to his theory of natural selection:
  - Each species produces more offspring than can survive.
  - Offspring compete with each other for limited resources.
  - Organisms in every population vary.
  - The offspring with the most favorable traits are most likely to survive and reproduce.

- Evidence for evolution includes:
  - fossils
  - biogeography
  - comparison of developmental embryology
  - comparative anatomy, including homologous and analogous structures
  - molecular biology (sequences of genes are conserved across many types of species)

- Members of a species are defined by the ability to reproduce to produce fertile offspring. Evolution and speciation occur when environmental pressures favor traits that permit survival and reproduction of selected individuals in a varied population. The favored traits may evolve to cause divergent or convergent evolution.

- The Hardy-Weinberg equations can be used to determine genetic variation within a population. These equations are as follows:
  - $p + q = 1$      Frequency of the dominant ($p$) and recessive ($q$) alleles
  - $p^2 + 2pq + q^2 = 1$      Frequency of the homozygous dominants ($p^2$), heterozygotes ($2pq$), and homozygous recessives ($q^2$)

- The Hardy-Weinberg equations describe a population that is not evolving but is instead said to be in genetic equilibrium. Such a population will only be so if it is not violating any of the following five conditions (almost all populations are evolving due to a violation of one or more of these factors):
  - large population
  - no mutations
  - no immigration or emigration
  - random mating
  - no natural selection

# Chapter 12 Drill

Answers and explanations can be found in Chapter 16.

1. The eye structures of mammals and cephalopods such as squid evolved independently to perform very similar functions and have similar structures. This evolution is an example of which of the following?

   (A) Allopatric speciation
   (B) Sympatric association
   (C) Divergent evolution
   (D) Convergent evolution

Questions 2–5 refer to the following graph and paragraph.

During the industrial revolution, a major change was observed in many insect species due to the mass production and deposition of ash and soot around cities and factories. One of the most famous instances was within the spotted moth population. An ecological survey was performed where the number of spotted moths and longtail moths were counted in 8 different urban settings over a square kilometer in 1802. A repeat experiment was performed 100 years later in 1902. The results of the experiment are shown below.

2. What type of selection is represented by the results of this study?

   (A) Stabilizing selection
   (B) Directional selection
   (C) Disruptive selection
   (D) Divergent selection

3. Which of the following statements best explains the data?

   (A) As time passed from 1802 to 1902, the frequency of white-bodied pepper moths increased and black-bodied pepper moths decreased.
   (B) As time passed from 1802 to 1902, the frequency of white-bodied pepper moths decreased and black-bodied pepper moths increased.
   (C) As time passed from 1802 to 1902, the frequency of white-bodied pepper moths and black-bodied pepper moths both increased.
   (D) As time passed from 1802 to 1902, the frequency of white-bodied pepper moths and black-bodied pepper moths both decreased.

4. Why was the population of longtail moths also surveyed in this study?

   (A) Variations in the environment were expected to alter the population of longtail moths.
   (B) Longtail moths were included as a control because they were not expected to change appreciably due to changes associated with the Industrial Revolution.
   (C) As time passed from 1802 to 1902, the frequency of white-bodied pepper moths and black-bodied pepper moths both increased.
   (D) As time passed from 1802 to 1902, the frequency of white-bodied pepper moths and black-bodied pepper moths both decreased.

5. How would the results of this study have been different if factories produced white or light gray ash and soot rather than black?

   (A) There would be no change to the results of the experiment.
   (B) There would been added selection pressure for more white-bodied spotted moths and against black-bodied spotted moths.
   (C) There would been added selection pressure for more black-bodied spotted moths and against white-bodied spotted moths.
   (D) There would have been an increase in the frequency of both black-bodied and white-bodied spotted moths.

Question 6 represents a question requiring a numeric answer.

6. A recessive allele of a gene has a calculated frequency of 0.3 in a population. Assuming the population is in Hardy-Weinberg equilibrium, what percentage of the population is expected to be heterozygous for the gene? Give your answer to the nearest hundredths place.

Questions 8–9 refer to the following scenario.

A cruise ship with 500 people on board crashes on a deserted island. They like the island so much that they decide to stay. There is a small number of people on board with red hair, an autosomal recessive trait.

8. What is most likely to occur with respect to the number of red-haired individuals on the island?

(A) There will be no change in the overall number of red-haired people in the population.
(B) Convergent evolution will occur, with all people eventually becoming red-haired.
(C) It is likely that over many generations, the proportion of red-haired people in the population will increase.
(D) Red-haired people will probably decrease in the population over many generations.

9. Evolution does not occur when a population is said to be in equilibrium by the Hardy-Weinberg definition. In the situation above, which of the following does not support the claim that evolution will occur in this population?

(A) The situation describes a small population.
(B) Mutations occur in this population.
(C) People do not usually mate randomly.
(D) Natural selection pressures do not exist on this island.

7. The Middle East blind mole-rat (*Nannosplalax ehrenbergi*) lives in the Upper Galilee Mountains of Israel. Two groups of these mole-rats live in the same region, but scientists discovered that there is a 40% difference in mitochondrial DNA between these two groups. Due to soil differences in these mountains, different plants grow in different regions of the mountains, which seems to coincide with the delineation between the mole-rat groups. These rats do not seem to interbreed in the wild. This is an example of

(A) allopatric speciation
(B) sympatric speciation
(C) prezygotic barrier
(D) hybrid zone

10. Peacock males have evolved to have huge, beautiful tails with numerous large eye spots. Long, colorful tails are difficult to carry when you are running from a predator. However, peahens (female peacocks) have been shown by researchers to mate preferably with males with more eye spots and longer tails, despite these traits making the males more susceptible to predation. Over time, this preference has resulted in peacocks with huge tails due to

(A) sexual selection
(B) disrupted selection
(C) divergent evolution
(D) sympatric speciation

# REFLECT

Respond to the following questions:

- For what content topics discussed in this chapter do you feel you have achieved sufficient mastery to answer multiple-choice questions correctly?

- For which content topics discussed in this chapter do you feel you have achieved sufficient mastery to discuss effectively in a free response?

- For which content topics discussed in this chapter do you feel you need more work before you can answer multiple-choice questions correctly?

- For which content topics discussed in this chapter do you feel you need more work before you can discuss effectively in a free response?

- What parts of this chapter are you going to re-review?

- Will you seek further help, outside this book (such as a teacher, tutor, or AP Central), on any of the content in this chapter—and if so, on what content?

# Chapter 13
# Behavior and
# Ecology

In the preceding chapters, we looked at individual organisms and the ways in which they solve life's many problems: acquiring nutrition, reproducing, and so on. Now let's turn to how organisms deal with their environments. We can divide the discussion of organisms and their environments into two general categories: **behavior** and **ecology**.

## BEHAVIOR

Some animals behave in a programmed way to specific stimuli, while others behave according to some type of learning. We humans can do both. For the AP Biology Exam, you'll have to know a little about the different types of behavior.

All behavior means, basically, is how organisms cope with their environments. This chapter looks at the general types of behavior: instinct, imprinting, classical conditioning, operant conditioning, and insight. Let's start with instinct.

## INSTINCTS

**Instinct** is an inborn, unlearned behavior. Sometimes the instinctive behavior is triggered by environmental signals called releasers. The releaser is usually a small part of the environment that is perceived. For example, when a male European robin sees another male robin, the sight of a tuft of red feathers on the male is a releaser that triggers fighting behavior. In fact, because instinct underlies all other behavior, it can be thought of as the circuitry that guides behavior.

For example, hive insects, such as bees and termites, never learn their roles; they are born knowing them. On the basis of this inborn knowledge, or instinct, they carry out all their other behaviors: A worker carries out "worker tasks," a drone "drone tasks," and a queen "queen tasks." Another example is the "dance" of the honey bee, which is used to communicate the location of food to other members of the beehive.

There are other types of instinct that last for only a part of an animal's life and are gradually replaced by "learned" behavior. For example, human infants are born with an ability to suck from a nipple. If it were not for this instinctual behavior, the infant would starve. Ultimately, however, the infant will move beyond this instinct and learn to feed itself. What exactly, then, is instinct?

For our purposes:

Instinct is the inherited "circuitry" that directs and guides behavior.

A particular type of innate behavior is a **fixed action pattern**. These behaviors are not simple reflexes, and yet they are not conscious decisions. An example is the egg-rolling behavior exhibited by a graylag goose. If the egg is removed from the goose, it will continue to make the same movements. That is, the innate movements are independent of the environment.

# Learning

Another form of behavior is **learning**. Learning refers to a change in a behavior brought about by an experience (which is what you're doing this very moment). Animals learn in a number of ways. For the AP Biology Exam, we'll take a look at the key types of learning.

# Imprinting

Have you ever seen a group of goslings waddling along after their mother? How is it that they recognize her? Well, the mother arrives and gives out a call that the goslings "recognize." The goslings, hearing the call, know that this is their mother and follow her around until they are big enough to head out on their own.

Now imagine the same goslings, newly hatched. If the mother is absent, they will accept the first moving object they see as their mother. This process is known as **imprinting**.

Animals undergo imprinting within a few days after birth in order to recognize members of their own species. While there are different types of imprinting—including parent, sexual, and song imprinting—they all occur during a **critical period**—a window of time when the animal is sensitive to certain aspects of the environment.

Remember:

> Imprinting is a form of learning that occurs during a brief period of time, usually early in an organism's life.

# Classical Conditioning

If you have a dog or a cat, you know that every time you hit the electric can opener, your cat or dog comes running. This is a form of **classical conditioning**. To feed your pet, you need to open its can of food. For your pet, the sound of the opener has come to be associated with eating: Every time it hears the opener, it thinks that it is about to be fed. We can say that your pet has been "conditioned" to link the buzz of the can opener and its evening meal.

The classic experiment demonstrating conditioning was done by a Russian scientist named Ivan Pavlov, who made his dogs salivate by ringing a bell. He did this the same way you make your dog come running with the can opener. Each time he fed his dogs, Pavlov rang a bell. Eventually, the dogs came to "associate" the bell with the food. By the end of the experiment, Pavlov had merely to ring the bell to start the dogs salivating. This type of learning is now known as **associative learning**.

Pavlov's experiments demonstrated what many of us know already: We can learn through conditioning, or repeated instances of an event.

## Operant Conditioning

Another type of associative learning is **operant conditioning** (or **trial-and-error learning**).

In operant conditioning, an animal learns to perform an act in order to receive a reward. This type of behavior was extensively studied by psychologist B. F. Skinner. He put a rat in an experimental cage and watched to see if it would randomly press different levers. Through trial and error, the animal figured out that one lever in particular would always produce food from a dispenser. Over time, the animal made an "association" between pressing the lever and getting food (the reward). Skinner even detected that some rats were so "conditioned" that they started to "hang out" near the lever. These same rats were also subjected to a negative form of operant conditioning, in which touching the bar led to an electric shock. As you could imagine, rats quite quickly learned not to touch the bar that gave them the shock.

If the behavior is not reinforced, the conditioned response will be lost. This is called extinction. Here's one thing to remember:

> In operant conditioning, the animal's behavior determines whether it gets the reward or the punishment.

**Habituation** is another form of learning. It occurs when an animal learns not to respond to a stimulus. For example, if an animal encounters a stimulus over and over again without any consequences, the response to it will gradually lessen and may altogether disappear. One example is how a marine worm, *Nerels*, withdraws into its protective tube if a shadow passes over. With repeated exposure to the stimulus, however, the response decreases.

## Insight

This is the highest form of learning and is exercised only by higher animals. **Insight** means the ability to figure out a behavior that generates a desired outcome. It is sometimes referred to as the "aha experience." As far as the animal world is concerned, human beings tend to be pretty good at using insight, or **reasoning**, to solve problems. However, we are not the only ones who reason. Chimpanzees, for example, have been known to use rudimentary tools, such as twigs and stones, to get their food.

To recap, there are four basic types of learning:

- Imprinting occurs early in life and helps organisms recognize members of their own species.

- Classical conditioning involves learning through association.

- Operant conditioning occurs when a response is associated with new stimuli (also a form of associative learning).

- Insight involves "reasoning" or problem solving.

## Internal Clocks: The Circadian Rhythm

There are other instinctual behaviors that occur in both animals and plants. One such behavior deals with time. Have you ever wondered how roosters always know when to start crowing? The first thought that comes to mind is that they've caught a glimpse of the sun. Yet many crow even before the sun has risen.

Roosters do have internal alarm clocks. Plants have them as well. These internal clocks, or cycles, are known as **circadian rhythms**.

Circadian rhythms are yet another example of instinct. When an organism does something on a daily basis, we say it acts according to its circadian rhythm. But how do we know for certain that it's instinctual?

In a famous experiment, an American scientist took a bunch of plants and animals to the South Pole and put them on a turntable set to rotate at exactly the same speed as the earth but in the opposite direction of the earth's rotation. As a result, the organisms had absolutely no indication of day or night. Yet all of them continued to carry out their regular 24-hour cycles. This proved that the cycles have nothing to do with sunlight and everything to do with the internal clock.

### The Science Behind Your Jet Lag
If you've ever flown overseas, you know all about circadian rhythms. They're the basis of jet lag. Our bodies tell us it's one time, while our watches tell us it's another. The sun may be up, but our body's internal clock is crying "Sleep!" This sense of time is purely instinctual: You don't need to know how to tell time in order to feel jet lag.

Watch out though: Seasonal changes, like the loss of leaves by deciduous trees or the hibernation of mammals, are not examples of circadian rhythms. *Circadian* refers only to daily rhythms. Need a mnemonic? Just think how bad your jet lag would be after a trip around the world. In other words: Circling the globe screws up your circadian rhythm.

# HOW ANIMALS COMMUNICATE

Some animals use signals as a way of communicating with members of their species. These signals, which can be chemical, visual, electrical or tactile, are often used to influence mating and social behavior.

Chemical signals are one of the most common forms of communication among animals. **Pheromones**, for instance, are chemical signals between members of the same species that stimulate olfactory receptors and ultimately affect behavior. For example, when female insects give off their pheromones, they attract males from great distances.

Visual signals also play an important role in the behavior observed among members of a species. For example, fireflies produce pulsed flashes that can be seen by other fireflies far away. The flashes are sexual displays that help male and female fireflies identify and locate each other in the dark.

Other animals use electrical channels to communicate. For example, some fish generate and receive weak electrical fields. Finally, tactile signals are found in animals that have mechanoreceptors in their skin to detect prey. For instance, cave-dwelling fishes use mechanoreceptors in their skin for communicating with other members and detecting prey.

## Social Behavior

Many animals are highly social species, and they interact with each other in complex ways. Social behaviors can help members of the species survive and reproduce more successfully. Several behavioral patterns for animal societies are summarized below:

- **Agonistic behavior** is aggressive behavior that occurs as a result of competition for food or other resources. Animals will show aggression toward other members that tend to use the same resources. A typical form of aggression is fighting between competitors.

- **Dominance hierarchies** (or pecking orders) occur when members in a group have established which members are the most dominant. The more dominant male will often become the leader of the group and will usually have the best pickings of the food and females in the group. Once the dominance hierarchy is established, competition and tension within the group is reduced.

- **Territoriality** is a common behavior when food and nesting sites are in short supply. Usually, the male of the species will establish and defend his territory (called a home range) within a group in order to protect important resources. This behavior is typically found among birds.

- **Altruistic behavior** is defined as unselfish behavior that benefits another organism in the group at the individual's expense because it advances the genes of the group. For example, when ground squirrels give warning calls to alert other squirrels of the presence of a predator, the calling squirrel puts itself at risk of being found by the predator.

## PLANT BEHAVIOR

Plants have also evolved specific ways to respond to their environment. The plant behaviors covered on the AP Biology Exam are photoperiodism and tropisms. Plants flower in response to changes in the amount of daylight and darkness. This is called **photoperiodism**. Although you'd think that plants bloom based on the amount of sunlight they receive, they actually flower according to the amount of uninterrupted darkness.

## Tropisms

Plants need light. Notice that all the plants in your house tip toward the windows. This movement of plants toward the light is known as phototropism. As you know, plants generally grow up and down: The branches grow upward, while the roots grow downward into the soil, seeking water. This tendency to grow toward or away from the earth is called gravitropism. All of these tropisms are examples of behavior in plants.

A **tropism** is a turning in response to a stimulus.

There are three basic tropisms in plants. They're easy to remember because their prefixes the stimuli to which plants react:

- **Phototropism** refers to how plants respond to sunlight. For example, bending toward light.

- **Gravitropism** refers to how plants respond to gravity. Stems exhibit negative gravitropism (they grow up, away from the pull of gravity), whereas roots exhibit positive gravitropism (they grow downward into the earth).

- **Thigmotropism** refers to how plants respond to touch. For example, ivy grows around a post or trellis.

# ECOLOGY

The study of the interactions between living things and their environments is known as ecology. We've spent most of our time discussing individual organisms. However, in the "real world," organisms are in constant interaction with other organisms and the environment. The best way for us to understand the various levels of ecology is to progress from the big picture, the biosphere, down to the smallest ecological unit, the population.

Just as in taxonomy, there is a hierarchy within the "ecology world." Each of the following terms represents a different level of ecological interaction.

- **Biosphere**—The entire part of the earth where living things exist. This includes soil, water, light, and air. In comparison to the overall mass of the earth, the biosphere is relatively small. If you think of the earth as a basketball, the biosphere is equivalent to a coat of paint over its surface.

- **Ecosystem**—The interaction of living and nonliving things.

- **Community**—A group of populations interacting in the same area.

- **Population**—A group of individuals that belong to the same species and that are interbreeding.

# Biosphere

The biosphere can be divided into large regions called **biomes**. Biomes are massive areas that are classified mostly on the basis of their climates and plant life. Because of the different climates and terrains on the earth, the distribution of living organisms varies. For the AP Biology Exam, you're expected to know both the names of the different biomes and their characteristic flora (plant life) and fauna (animal life).

Here's a summary of the major biomes:

---

## MAJOR BIOMES

**Tundra**
*Regions*—northernmost regions
*Plant life*—few, if any, trees; primarily grasses and wildflowers
*Characteristics*—contains permafrost (a layer of permanently frozen soil); has a short growing season
*Animal life*—includes lemmings, arctic foxes, snowy owls, caribou, and reindeer

**Taiga**
*Region*—northern forests
*Plant life*—wind-blown conifers (evergreens), stunted in growth, possess modified spikes for leaves
*Characteristics*—very cold, long winters
*Animal life*—includes caribou, wolves, moose, bear, rabbits, and lynx

**Temperate Deciduous Forest**
*Regions*—northeast and middle eastern United States, western Europe
*Plant life*—deciduous trees that drop their leaves in winter
*Characteristics*—moderate precipitation; warm summers, cold winters
*Animal life*—includes deer, wolves, bear, small mammals, birds

**Grasslands**
*Regions*— American Midwest, Eurasia, Africa, South America
*Plant life*—grasses
*Characteristics*—hot summers, cold winters; unpredictable rainfall
*Animal life*—includes prairie dogs, bison, foxes, ferrets, grouse, snakes, and lizards

**Deserts**
*Regions*—western United States
*Plant life*—sparse, includes cacti and drought-resistant plants
*Characteristics*—arid, low rainfall; extreme diurnal temperature shifts
*Animal life*—includes jackrabbits (in North America), owls, kangaroo rats, lizards, snakes, tortoises

**Tropical Rain Forests**
*Regions*—South America
*Plant life*—high biomass; diverse types
*Characteristics*—high rainfall and temperatures; impoverished soil
*Animal life*—includes sloths, snakes, monkeys, birds, leopards, and insects

---

Remember that the biomes tend to be arranged along particular latitudes. For instance, if you hiked from Alaska to Kansas, you would pass through the following biomes: tundra, taiga, temperate deciduous forests, and grasslands.

## Ecosystem

Ecosystems are self-contained regions that include both living and nonliving factors. For example, a lake, its surrounding forest, the atmosphere above it, and all the organisms that live in or feed off the lake would be considered an ecosystem. As you probably know, there is an exchange of materials between the components of an ecosystem. Take a look at the flow of carbon through a typical ecosystem:

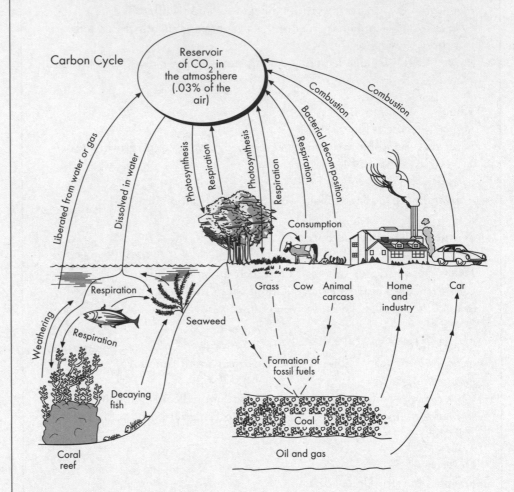

You'll notice how carbon is recycled throughout the ecosystem—this is called the **carbon cycle**. In other words, carbon flows through ecosystems. Other substances have similar cycles.

## Community

The next smaller level is the community. A community refers to a group of interacting plants and animals that show some degree of interdependence. For instance, you, your dog, and the fleas on your dog are all members of the same community. All organisms within a community fill one of the following roles: producers (or autotrophs), consumers (or heterotrophs), or decomposers.

## Producers

Producers, or autotrophs, have all of the raw building blocks to make their own food. From water and the gases that abound in the atmosphere, and with the aid of the sun's energy, autotrophs convert light energy to chemical energy. They accomplish this through photosynthesis.

## Consumers

Consumers, or heterotrophs, are forced to find their energy sources in the outside world. Basically, heterotrophs digest the carbohydrates of their prey into carbon, hydrogen, and oxygen, and use these molecules to make organic substances.

The bottom line is: Heterotrophs, or consumers, get their energy from the things they consume.

## Decomposers

All organisms at some point must finally yield to decomposers. Decomposers are the organisms that break down organic matter into simple products. Generally, fungi and bacteria are the decomposers. They serve as the "garbage collectors" in our environment.

## Niche

Each organism has its own **niche**—its position or function in a community. Because every species occupies a niche, it's going to have an impact on all the other organisms. These connections are shown in the **food chain**. A food chain describes the way different organisms depend on one another for food. There are basically four levels to the food chain: producers, primary consumers, secondary consumers, and tertiary consumers.

Autotrophs produce all of the available food. They make up the first trophic (feeding) level. They possess the highest biomass (the total weight of all the organisms in an area) and the greatest numbers. Did you know that plants make up about 99 percent of the earth's total biomass?

**Primary consumers** are organisms that directly feed on producers. A good example is a cow. These organisms are also known as **herbivores**. They make up the second trophic level.

The next level consists of organisms that feed on primary consumers. They are the **secondary consumers**, and they make up the third trophic level. Above these are **tertiary consumers**.

So now you have it. We've got our four complete levels of the food chain.

> - Producers make their own food.
>
> - Primary consumers (herbivores) eat producers.
>
> - Secondary consumers (carnivores and omnivores) eat producers and primary consumers.
>
> - Tertiary consumers eat all of the above.

## The 10% Rule

In a food chain, only about 10 percent of the energy is transferred from one level to the next—this is called the **10% rule**. The other 90 percent is used for things like respiration, digestion, running away from predators—in other words, it's used to power the organism doing the eating! The producers have the most energy in an ecosystem; the primary consumers have less energy than producers; the secondary consumers have less energy than the primary consumers; and the tertiary consumers will have the least energy of all.

The energy flow, biomass, and numbers of members within an ecosystem can be represented in an **ecological pyramid**. Organisms that are "higher up" on the pyramid have less biomass and energy, and they are fewer in number.

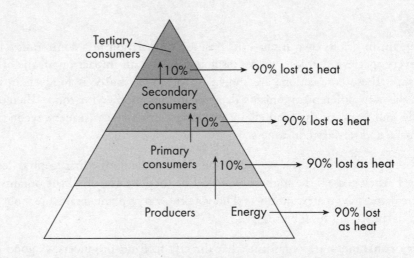

So, for example, primary consumers have less biomass and are fewer in number than producers. But what about decomposers? Where do they fit in on the food chain? They don't. They are not really considered part of the food chain. Decomposers are usually placed just below the food chain to show that they can decompose any organism.

In addition to the distinctions drawn above, remember:

> Toxins in an ecosystem are more concentrated and thus more dangerous for animals further up the pyramid.

This simply means that if a toxin is introduced into an ecosystem, the animals most likely to be affected are those at the top of the pyramid. This occurs because of the increasing concentration of such toxins. The classic example of this phenomenon is DDT, an insecticide initially used to kill mosquitoes. Although the large-scale spraying of DDT resulted in a decrease in the mosquito population, it also wound up killing off ospreys.

Ospreys are aquatic birds whose diet consists primarily of fish. The fish that ospreys consumed had, in turn, been feeding on contaminated insects (bioaccumulation). Because fish eat thousands of insects, and ospreys hundreds of fish, the toxins grew increasingly concentrated (biomagnification). Though the insecticide seemed harmless enough, it resulted in the near-extinction of certain osprey populations. What no one knew was that in sufficient concentrations, DDT weakened the eggshells of ospreys. Consequently, eggs broke before they could hatch, killing the unborn ospreys.

Such environmental tragedies still occur. All ecosystems, small or great, are intricately woven, and any change in one level invariably results in major changes at all the other levels.

**Ecosystem Interconnectedness**
Because all levels of an ecosystem are closely connected, a disruption in one level due to environmental changes, for example, affects all other aspects of that ecosystem.

## Symbiotic Relationships

Many organisms that coexist exhibit some type of symbiotic relationship. These include remoras, or "sucker fish," which attach themselves to the backs of sharks, and lichen—the fuzzy, mold-like stuff that grows on rocks. Lichen appears to be one organism, when in fact it is two organisms—a fungus and an alga, or photosynthetic bacterium—living in a complex symbiotic relationship.

Overall, there are three basic types of symbiotic relationships:

- **Mutualism**—in which both organisms win (for example, the lichen).

- **Commensalism**—in which one organism lives off another with no harm to the "host" organism (for example, the remora).

- **Parasitism**—in which the organism actually harms its host.

# POPULATION ECOLOGY

Population ecology is the study of how populations change. Whether these changes are long-term or short-term, predictable or unpredictable, we're talking about the growth and distribution patterns of a population.

When studying a population, you need to examine four things: the size (the total number of individuals), the density (the number of individuals per area), the distribution patterns (how individuals in a population are spread out), and the age structure.

One way to understand the growth pattern and make predictions about the population growth of a country is by examining age structure histograms. For example, in underdeveloped countries, where the population increase is high, the base of the histogram is very wide compared to countries that show moderate growth.

Another way to study the changes in a population is by looking at survivorship curves. These curves are graphs of the numbers of individuals surviving to different ages, indicating the probability of any individual living to a given age.

For example, the graph below shows there is a high death rate among the young of oysters, but those that survive do well. On the other hand, there is a low death rate among the young of humans, but, after age 60, the death rate is high.

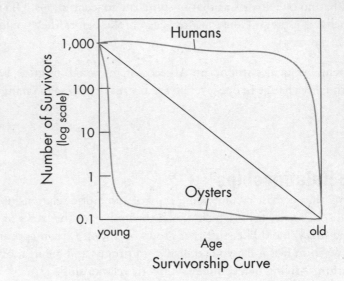

Survivorship Curve

The growth of a population can be represented as the number of crude births minus the number of crude deaths divided by the size of the population,

$$r = (births - deaths)/N$$

($r$ is the reproductive rate, and $N$ is the population size)

Each population has a **carrying capacity**—the maximum number of individuals of a species that a habitat can support. Most populations, however, don't reach their carrying capacity because they're exposed to limiting factors.

One important factor is **population density**. The factors that limit a population are either density-independent or density-dependent. **Density-independent factors** are factors that affect the population, regardless of the density of the population. Some examples are severe storms and extreme climates. On the other hand, **density-dependent factors** are those with effects that depend on population density. Resource depletion, competition, and predation are all examples of density-dependent factors. In fact, these effects become even more intense as the population density increases.

# Exponential Growth

The growth rates of populations also vary greatly. There are two types of growth: exponential growth and logistic growth. **Exponential growth** occurs when a population is in an ideal environment. Growth is unrestricted because there are lots of resources, space, and no disease or predation. Here's an example of exponential growth. Notice that the curve arches sharply upward—the exponential increase.

Exponential growth occurs very quickly, resulting in a J-shaped curve. A good example of exponential growth is the initial growth of bacteria in a culture. There's plenty of room and food, so they multiply rapidly—every 20 minutes.

However, as the bacterial population increases, the individual bacteria begin to compete with each other for resources. The population reaches its carrying capacity, and the curve levels off.

The population becomes restricted in size because of limited resources. This is referred to as **logistic growth**. Notice that the growth forms an S-shaped curve. These growth patterns are associated with two kinds of life-history strategies: r-selected species and k-selected species.

We've already mentioned that organisms that grow exponentially approach the carrying capacity. These organisms tend to thrive in areas that are barren or uninhabited. Once they colonize an area, they reproduce as quickly as possible. Why? They know they've got to multiply before competitors arrive on the scene! The best way to ensure their survival is to produce lots of offspring. These organisms are known as **r-strategists**. Typical examples are common weeds, dandelions, and bacteria.

At the other end of the spectrum are the **k-strategists**. These organisms are best suited for survival in stable environments. They tend to be large animals, such as elephants, with long lifespans. Unlike r-strategists, they produce few offspring. Given their size, k-strategists usually don't have to contend with competition from other organisms.

## ECOLOGICAL SUCCESSION

Communities of organisms don't just spring up on their own; they develop gradually over time. **Ecological succession** refers to the predictable procession of plant communities over a relatively short period of time (decades or centuries).

Centuries may not seem like a short time to us, but if you consider the enormous stretches of time over which evolution occurs, hundreds of thousands or even millions of years, you'll see that it is pretty short.

The process of ecological succession in which no previous organisms have existed is called **primary succession**.

## The Pioneers

How does a new habitat full of bare rocks eventually turn into a forest? The first stage of the job usually falls to a community of lichens. Lichens are hardy organisms. They can invade an area, land on bare rocks and erode the rock surface, and over time turn it into soil. Lichens are considered **pioneer organisms**.

Once lichens have made an area more habitable, they've set the stage for other organisms to settle in. Communities establish themselves in an orderly fashion. Lichens are replaced by mosses and ferns, which in turn are replaced by tough grasses, then low shrubs, then evergreen trees, and finally, the deciduous trees. Why are lichens replaced? Because they can't compete with the new plants for sunlight and minerals.

Mineral sere · Grass-Herb sere · Shrub sere · White pine-Spruce sere · Beech-Maple sere

1 year   25 years   50 years   75 years   100 years

The entire sequence is called a **sere**. The final community is called the **climax community**. The climax community is the most stable. In our example, the deciduous trees are part of the climax community.

Now what happens when a forest is devastated by fire? The same principles apply, but the events occur much more rapidly. The only exception is that the first invaders are usually not lichens but grasses, shrubs, saplings, and weeds. When a new community develops where another community has been destroyed or disrupted, this event is called **secondary succession**.

## HUMAN IMPACT ON THE ENVIRONMENT

Unfortunately, humans have disturbed the existing ecological balance, and the results are far-reaching. Soils have been eroded and various forms of pollution have increased. The potential consequences on the environment are summarized below:

- **Greenhouse effect**—The increasing atmospheric concentrations of carbon dioxide through the burning of fossil fuels and forests have contributed to the warming of the earth. Higher temperatures may cause the polar ice caps to melt and flooding to occur. Other potential effects of global warming include changes in precipitation patterns, changes in plant and animal populations, and detrimental changes in agriculture.

- **Ozone depletion**—Pollution has also led to the depletion of the atmospheric ozone layer by such chemicals as chlorofluorocarbons (CFCs), which are used in aerosol cans. Ozone ($O_3$), forms when UV radiation reacts with $O_2$. Ozone protects the earth's surface from excessive ultraviolet radiation. Its loss could have major genetic effects and could increase the incidence of cancer.

- **Acid rain**—The burning of fossil fuels produces pollutants such as sulfur dioxide and nitrogen dioxide. When these compounds react with droplets of atmospheric water in clouds, they form sulfuric and nitric acids, respectively. The rain that falls from these clouds is weakly acid and is called acid rain. Acid rain lowers the pH of aquatic ecosystems and soil which damages water systems, plants, and soil. For example, the change in soil pH causes calcium and other nutrients to leach out, which damages plant roots and stunts their growth. Furthermore, useful microorganisms that release nutrients from decaying organic matter into the soil are also killed, resulting in less nutrients being available for the plants. Low pH also kills fish, especially those that have just hatched.

Acid rain, deforestation, pollution, soil erosion, and depletion of the ozone layer are just a few of the problems plaguing the environment today.

- **Desertification**—When land is overgrazed by animals, it turns grasslands into deserts and reduces the available habitats for organisms.

- **Deforestation**—When forests are cleared (especially by the slash and burn method), erosion, floods, and changes in weather patterns can occur.

- **Pollution**—Another environmental concern is the toxic chemicals in our environment. One example is DDT, a pesticide used to control insects. DDT was overused at one time and later found to damage plants and animals worldwide. DDT is particularly harmful because it resists chemical breakdown, and it can still be found in the tissues of nearly every living organism. The danger with toxins such as DDT is that as each trophic level consumes DDT, the substance becomes more concentrated by a process called **biomagnification**.

- **Reduction in biodiversity**—As different habitats have been destroyed, many plants and animals have become extinct. Some of these plants could have provided us with medicines and products that may have been beneficial.

## KEY TERMS

behavior
ecology
instinct
fixed action pattern
learning
imprinting
critical period
classical conditioning
associative learning
operant conditioning
  (or trial-and-error learning)
habituation
insight
reasoning
circadian rhythms
pheromones
agonistic behavior
dominance hierarchy
territoriality
altruistic behavior
photoperiodism
tropism
phototropism
gravitropism
thigmotropism
biosphere
ecosystem
community
population
biomes
carbon cycle
niche

food chain
primary consumers
herbivores
secondary consumers
tertiary consumers
10% rule
ecological pyramid
mutualism
commensalism
parasitism
carrying capacity
population density
density-independent factors
density-dependent factors
exponential growth
logistic growth
r-strategists
k-strategists
ecological succession
primary succession
pioneer organisms
sere
climax community
secondary succession
greenhouse effect
ozone depletion
acid rain
desertification
deforestation
pollution
biomagnification

# Summary

### Behavior

- Behavior is an organism's response to the environment. Behavior can be instinctual (inborn), learned, or social.
  - Types of learned behaviors include imprinting, classical conditioning and operant conditioning, and insight.
  - Types of social behaviors include agnostic behavior, dominance hierarchies, territoriality, and altruistic behavior.

- There are also plant-specific behaviors known as tropisms. The three basic tropisms are phototropism, gravitropism, and thigmotropism.

### Ecology

- There are several major biomes (tundra, taiga, temperate deciduous forest, grasslands, deserts, tropical rainforests) that make up the biosphere. Each biosphere contains ecosystems.

- Within an ecosystem are communities, which consist of organisms fulfilling one of three main roles:
  - Producers, or autotrophs, convert light energy to chemical energy via photosynthesis.
  - Consumers, or heterotrophs, acquire energy from the things they consume. Their digestion of carbohydrates produces carbon, hydrogen, and oxygen, which are then used to make organic substances.
  - Decomposers form fossil fuels from the detritus of other organisms in the ecosystem.

- The smallest unit of ecology is the population. The growth of a population can be found with the equation $(r) = (\text{births} - \text{deaths}) / N$.

- The carrying capacity is the maximum number of individuals that can be supported by a habitat. Most populations do not reach carrying capacity due to factors such as population density (density-independent factors and density-dependent factors).

- Exponential growth (J-shaped curve) occurs when a population is an ideal environment. Logistic growth (S-shaped curve) of a population occurs when there are limited resources in an environment.

- Organisms are generally either r-strategists or k-strategists. R-strategists ensure their survival by producing lots of offspring. K-strategists, which are usually large animals, produce few offspring but have longer life spans and less competition from other organisms for resources.

- Succession describes how ecosystems recover after a disturbance in terms of pioneers, sere, climax community, and secondary succession.

- Human impact on the planet includes issues such as:
  - greenhouse effect
  - ozone depletion
  - acid rain
  - desertification
  - deforestation
  - pollution
  - reduction in biodiversity

# Chapter 13 Drill

Answers and explanations can be found in Chapter 16.

1. A chimpanzee stacks a series of boxes on top of one another to reach a bunch of bananas suspended from the ceiling. This is an example of which of the following behaviors?

   (A) Operant learning
   (B) Imprinting
   (C) Instinct
   (D) Insight

2. Viruses are obligate intracellular parasites, requiring their host cells for replication. Consequently, viruses generally attempt to reproduce as efficiently and quickly as possible in a host. Below is a graph depicting the initial growth pattern of a bacteriophage within a population of *E. coli*. This reproductive strategy is most similar to which of the following?

   (A) An r-strategist, because it aims to produce a large abundance of offspring to ensure survival
   (B) A k-strategist, because it aims to produce a large abundance of offspring to ensure survival
   (C) An r-strategist, because it is best suited to thrive in stable environments and over a long life-span
   (D) A k-strategist, because it is best suited to thrive in stable environments and over a long life-span

3. In a pond ecosystem, spring rains trigger an expansion of species at levels of the food chain. Runoff from nearby hills brings nutrients which, when combined with warming temperatures, trigger an algae bloom. The populations of small protozoans such as plankton expand by ingesting the algae. These plankton, subsequently, are consumed by small crustaceans such as crayfish, which ultimately become prey for fish such as catfish or carp. In this ecosystem, which of the following accurately describes the crayfish?

   (A) They are producers.
   (B) They are primary consumers.
   (C) They are secondary consumers.
   (D) They are tertiary consumers.

Questions 4–6 refer to the following table, chart and paragraph.

**Table 1**: Minimum pH Tolerance of Common Aquatic Organisms

| Animal | pH Minimum |
|---|---|
| Brook Trout | 4.9 |
| American Bullfrog | 3.8 |
| Yellow Perch | 4.6 |
| Tiger Salamander | 4.9 |
| Crayfish | 5.4 |
| Snails | 6.1 |
| Clams | 6.0 |

**Chart 1**: Change in Populations of Aquatic Organisms in Richard Creek (1980 – 2000)

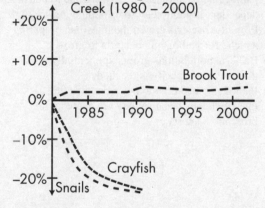

4. According to the data in Chart 1 and Table 1, the average pH in the creek is most nearly which of the following?

(A) 5.9
(B) 5.5
(C) 5.1
(D) 4.7

5. Which of the following is the most likely explanation for the abrupt decrease in pH in Richard Creek?

(A) Greenhouse effect
(B) Deforestation
(C) Acid rain
(D) Pollution

6. In order for snails to return to Richard Creek, the pH of the creek must exceed

(A) 5.4
(B) 5.7
(C) 6.0
(D) 6.1

Over the period of 1980 to 2000, the average pH in Richard Creek changed drastically. An ecological survey was performed to evaluate the effect of detectable decreases in pH on the aquatic life of the creek. Four times a year, ecological surveys were performed to identify the amount of snails (a primary consumer), crayfish (a secondary consumer), and brook trout (a tertiary consumer) present at five different locations. The percent change relative to 1980 is shown in **Chart 1** above. Many aquatic organisms cannot live in low pH conditions. The minimum pH necessary for common aquatic organisms to sustain life is shown in **Table 1**.

7. *Mimosa pudica* is a plant often called the "sensitive plant" because when you touch the leaves, they immediately close up. One theory about the purpose of this type of movement is that herbivores avoid the plant due to this movement. The movement of *Mimosa pudica* is an example of

(A) phototropism
(B) gravitropism
(C) aquatropism
(D) thigmotropism

Question 8 refers to the following graph.

## Foodprints by Diet Type: t $CO_2^e$/person

8. A proponent of the vegan diet writes a blog post that claims that the vegan diet is the only way to "reduce your carbon footprint, with respect to food, because it cuts your carbon footprint in half" and shows the graph above. Which of the following directly supports this claim?

(A) According to the 10% rule, only 10% of energy is passed from producers to consumers, so when humans consume animals, the energy harvested is greatly reduced.
(B) Livestock produce methane emissions and consume large amounts of plants every year.
(C) Deforestation for grazing livestock contributes to the loss of plants for carbon sequestration.
(D) Beef and lamb livestock—not chicken, veal, or pork—has the most dramatic impact on carbon footprint.

9. Costa Rica is plentiful with many beautiful tropical rainforests. For a long time, cattle farmers cleared much of the forests for their livestock. However, in recent decades, the government has placed more emphasis on reforestation. Several fields are now being reclaimed by the rainforest. One such field has many long grasses, a few tall palms, and some dense shrubbery along the periphery of the field. This field can be said to be

(A) invaded by pioneer species
(B) a product of secondary succession
(C) a climax community
(D) filled with invasive species

10. Probiotics are beneficial bacteria or yeast applied to maintain intestinal health in humans or livestock. In theory, these organisms inhabit the intestines and aid in digestion of food. Also, these organisms have bioprotective effects. How can addition of bacteria to the intestine protect the host from disease?

(A) Probiotic bacteria crowd the intestine, preventing pathogenic bacteria from growing due to density-dependent limitation on their population.
(B) Probiotic bacteria crowd the intestine, preparing a niche for pathogenic bacteria to grow.
(C) Bacteria populations grow exponentially until the carrying capacity is reached.
(D) Pathogenic bacteria do not recognize the intestinal cells because they are coated with probiotic bacteria.

# REFLECT

Respond to the following questions:

- For what content topics discussed in this chapter do you feel you have achieved sufficient mastery to answer multiple-choice questions correctly?

- For which content topics discussed in this chapter do you feel you have achieved sufficient mastery to discuss effectively in a free response?

- For which content topics discussed in this chapter do you feel you need more work before you can answer multiple-choice questions correctly?

- For which content topics discussed in this chapter do you feel you need more work before you can discuss effectively in a free response?

- What parts of this chapter are you going to re-review?

- Will you seek further help, outside this book (such as a teacher, tutor, or AP Central), on any of the content in this chapter—and if so, on what content?

# Chapter 14
# Sample Free-Response Questions

## SECTION II: FREE RESPONSE

Section II of the AP Biology Exam contains eight questions, consisting of two long free-response questions and six short free-response questions. You will have a total of 90 minutes—including 10 minutes for reading—to complete all questions. The directions will look something like this (read carefully!):

> **Directions:** Questions 1 and 2 are long free-response questions that should require about 22 minutes each to answer and are worth 10 points each. Questions 3 through 8 are short free-response questions that should require about 6 minutes each to answer. Questions 3 through 5 are worth 4 points each, and questions 6 through 8 are worth 3 points each.
>
> Read each question carefully and completely. Write your response in the space provided following each question. Only material written in the space provided will be scored. Answers must be written out in paragraph form. Outlines, bulleted lists, or diagrams alone are not acceptable.

The following sections discuss scoring guidelines in more detail and provide sample questions. But first take some time to review some key tips for writing your free responses.

## Free-Response Tips and Strategies

AP Biology readers (the people who grade your free-response essays) expect students to interpret data and apply knowledge to new situations in the free-response questions. Some important things to keep in mind when writing your answers to the free-response questions are provided in the following list.

- **Answer the question.** Do not become distracted or waste time writing about tangential topics that you know more about than what is being asked. Focus your response.

- **Think quantitatively.** Use data whenever possible to explain your answer. This is especially necessary when the question features a graph, table, or other type of diagram (which many free-response questions now do).

- **Be specific.** Offer specific detailed examples where appropriate.

- **Use vocabulary correctly and often.** Hone your understanding of the words provided in the Key Terms lists in this book and the various biological ideas and situations to which they can be applied. This will help you use the correct terminology when writing your responses.

- **Remember the Big Ideas.** Know the four major themes covered in the AP Biology course (page 49), as they are reflected in many free-response questions. Understanding these ideas and being able to articulate them will help you craft your responses.

Okay, now it's time for some practice. Take about 22 minutes and try to write a long free-response to the following question. Keep these tips in mind as you write.

1.  All organisms need nutrients to survive. Angiosperms and vertebrates have each developed various methods to obtain nutrients from their environment.

    (a)  **Discuss** the ways angiosperms and vertebrates procure their nutrients.
    (b)  **Discuss** two structures used for obtaining nutrients among angiosperms. **Relate** structure to function.
    (c)  **Discuss** two examples of symbiotic relationships that have evolved between organisms to obtain nutrients.

# SCORING GUIDELINES

To help you grade this sample free response, we've put together a checklist that you can use to calculate the number of points that should be assigned to each part of this question. We'll first explain the important points on the checklist, and then give you sample student responses to show you how test reviewers would evaluate them.

## Free-Response Checklist

### Part (a): Types of Nutritional Requirements—4 Points Maximum

**Angiosperms:**
**1 point**—Autotrophs—if defined
**1 point**—They require $H_2O$, $CO_2$, and minerals.

**Vertebrates:**
**1 point**—Heterotrophs—if defined
**1 point**—They require organic compounds (sugars, proteins, fats), $H_2O$, vitamins, and minerals.

## Part (b): Structures in Angiosperms—4 Points Maximum

First two structures only

| (1 point each) | (1 point each) |
| --- | --- |
| **Structure** | **Function described** |
| Cuticles | Prevent desiccation |
| Stomata | Pores in leaf that regulate gas exchange |
| Lenticels | Pores in stems that regulate gas exchange |
| Palisade mesophyll | Loosely packed cells in a leaf; contains chloroplasts |
| Roots | Structures that absorb water and minerals |
| Root hairs | Extensions of the root that increase the surface area for water absorption |

**(1 point)**

Carnivorous plants (such as Venus flytrap) have modified leaves to trap insects.

## Part (c): Symbiotic Relationships—4 Points Maximum

First 2 relationships only

| (1 point each) | (1 point each) |
| --- | --- |
| **Relationship** | **Benefits to Each** |
| Nitrogen-fixing bacteria | Bacteria get food/Plants get usable nitrogen |
| Mycorrhizae | Fungi get food/Plants get water and minerals |
| Intestinal bacteria | Humans get Vitamin K/Bacteria have a place to live |
| Ruminants flora | Bacteria digest the cellulose for cows/Bacteria have a place to live |
| Wood-eating termites | Protists digest wood/Termites obtain the wood |

# Explanation of Checklist

For Part (a), you'll receive points if you demonstrate a clear understanding of the terms autotroph *and* heterotroph. You must define the terms and describe the necessary nutrients for both angiosperms and vertebrates. For angiosperms, you have to mention the raw materials for photosynthesis ($CO_2$ + $H_2O$) and minerals. For vertebrates, you have to mention organic compounds and additional vitamins or minerals.

For Part (b), you have to relate structure to function. Which structures are involved in obtaining nutrients? You're asked to discuss *two* structures in angiosperms. One point is given for the structure, and another point is given if you show a clear understanding of the function of the structure.

For Part (c), you are asked to give *two* examples of symbiotic relationships. Since they ask for only two examples, you earn points only for the first two examples. An additional point is also given if you describe the benefits to each organism. Notice that you don't have to know everything about biology to get a high score on the essay section. However, you do have to present the proper information with a certain level of detail. Additionally, you can earn a maximum of 4 points for each part. However, you can't earn more than 10 points total.

Before you use this checklist to assign points to your response, let's evaluate two examples.

## SAMPLE ESSAYS

Let's read an essay written by an average student, which was given 4 points. (All grammar and spelling mistakes have been retained.)

## Sample Response 1

Angiosperms are autotrophs which make their own food. They have chloroplasts within the leaves which captures sunlight. Angiosperms also get nutrients through their roots which absorb water and bring it up the leaf. Food, water and minerals are then transport in the plant by xylem and phloem. Vertebrates can not make their own food so they must get their nutrients from other organisms. Some vertebrates only eat plants while other vertebrates eat both plants and animals.

Two structures that angiosperms use to obtain nutrients from the environment are leaves and roots. Plants have leaves which are shaped to absorb the maximum amount of sunlight. Plants also have root hairs which increase the surface area so that more water and minerals can be taken up the leaf.

An example of a symbiotic relationship is the bacteria found in the root nodules of plants. Plants need nitrogen to make plant proteins but they can't use the nitrogen from the atmosphere. Plants therefore work with nitrogen-fixing bacteria which convert nitrogen to nitrates. Both organisms benefit from this relationship.

## Explanation

This student would probably get a 4 for this essay. For Part (a), he clearly defines the term autotroph, which earns the essay 1 point. However, the student does not list the starting materials ($CO_2$ + $H_2O$) to make food in plants. He also fails to define heterotrophs. Consequently, he does not receive any points for his explanation of heterotrophs. Notice that for this paragraph, he is not penalized for any grammar mistakes.

For Part (b), the response receives 2 points for discussing root hairs and their function. Unfortunately, the explanation of leaves is not detailed enough, so the essay does not receive any additional points.

For Part (c), the response receives 1 point for the example of a symbiotic relationship—nitrogen-fixing bacteria and plants. However, the student does not specifically describe the benefits to *each* organism. Although he mentions the benefits to plants, he neglects to discuss the benefits to bacteria. Had the student completed this thought, he would have earned another point. Overall, although the essay is a little choppy, the student is not penalized for it.

Now let's read an essay written by another student that was given 9 points.

## Sample Response 2

Organisms have developed a variety of ways to obtain nutrients. Angiosperms, which are flowering plants, are autotrophs. This means that they are capable of making their own food via photosynthesis. The reaction for photosynthesis is:

$$6CO_2 + 12H_2O \xrightarrow{\text{sunlight}} C_6H_{12}O_6 + 6O_2 + 6H_2O + ATP$$

Plants use carbon dioxide and water to produce glucose and ATP. Angiosperms also use their roots to absorb water and essential minerals from the soil.

Vertebrates, however, are heterotrophs. They are unable to make their own food. They must therefore rely on other organisms to obtain energy. They require organic compounds, such as proteins and carbohydrates, and break them down to simpler molecules that can be absorbed into the body. Vertebrates also require other essential substances such as vitamins, which serve as coenzymes in crucial biological reactions.

Two structural adaptations used by angiosperms to obtain nutrients are root hairs and stomates. Plants contain numerous root hairs which are tiny projections from a plant's root that increase the surface area for the uptake of water and minerals. Plants also have stomates which regulate the intake of $CO_2$. The carbon dioxide that enters the stomates can then be used to make glucose.

There are many examples of symbiotic relationships. A common example is nitrogen fixation. Certain plants, called legumes, require nitrogen to synthesize plant amino acids. These plants have nitrogen-fixing bacteria which live in their root nodules and convert nitrogen to nitrates. The plant uses the nitrates to make plant proteins. The bacteria, in turn, gets energy from the carbohydrates of the plant.

Another example of a symbiotic relationship is bacteria in the stomach of cows. Cows are herbivores that can not digest cellulose. They rely on bacteria in their stomach to digest some of the cellulose found in their stomach. Cows, therefore, get some of their energy from fermenting bacteria. A third example of symbiosis is the bacteria found in vertebrates, such as humans. Humans have bacteria that reside in their large intestine. These microorganisms break down undigested food particles and produce Vitamin K. The bacteria in turn have a place to live.

## Explanation

This essay received 9 points. This student clearly demonstrates a mastery of the material. In Part (a), the student defines and discusses the terms autotroph and heterotroph. She also mentions the important raw materials that are involved in each case. Notice that she presents and explains the equation for photosynthesis. She earns 4 points—the maximum—for this section.

In Part (b), the student presents two structures and their function. Her coverage of the material is fairly thorough. She addresses two important structures: root hairs and stomates. She receives 2 points: 1 for mentioning each structure, and 1 point for the discussion of root hairs. She does not mention the structure of stomates—that they're pores in the epidermis—so she does not receive a point for the function of stomates.

For Part (c), the student presents three types of symbiotic relationships. The essay reader would probably review only the first two examples to assign points. The student receives 2 points for her discussion of nitrogen fixation but only 1 point for her explanation of bacteria in the cow's stomach. In this case, she forgets to mention the benefits to the bacteria.

# HOW TO USE THE SAMPLE FREE RESPONSES

The most important advice for this section of the test is *practice, practice, practice!* No matter how well you think you know the material, it's important to practice formulating your thoughts on paper. Before you take our practice tests in the back of the book, make sure you review our sample free-response questions. Then use our checklist—which is put together just like ETS's—to assign points.

To give you as much practice as possible, we've compiled a few sample long and short free-response questions. Most of these questions refer to themes in biology that crop up time and time again on the exam, so chances are you'll see something similar on your test. You can also access free-response questions from recent exams on the College Board website at **https://apstudent.collegeboard.org/apcourse/ap-biology/exam-practice.**

# SAMPLE LONG FREE-RESPONSE QUESTIONS

1. **Describe** how the modern theory of evolution is supported by evidence from the following areas:

   (a) Comparative anatomy
   (b) Biogeography
   (c) Embryology

2. **Describe** the chemical nature of genes. **Discuss** the replicative process of DNA in eukaryotic organisms. What are the various types of gene mutations that can occur during replication?

3. **Describe** the process of photosynthesis. What are the major plant pigments involved in photosynthesis? **Design** an experiment to measure the rate of photosynthesis.

# SAMPLE SHORT FREE-RESPONSE QUESTIONS

1. **Describe** the structure of a generalized eukaryotic cell.

2. **Discuss** the following responses in plants and give one example of each.

   (a) Photoperiodism
   (b) Phototropism

3. **Select** one of the following pairs of hormones and **discuss** the concept of negative feedback.

   (a) Thyroid-stimulating hormone (TSH) and thyroxine
   (b) Parathyroid hormone and calcitonin
   (c) Cortisol and ACTH

The biology knowledge you need to answer all of these free-response questions is in this book. After you have completed your essays, go back to the chapters that cover these topics and check to see if you have presented all the relevant points. Use our Free-Response Checklist as a guideline to give yourself (or have someone give you) an approximate score. Good luck!

# Summary

- The free-response section features eight questions, two of which require long responses. You will have 90 minutes to complete the entire section.

- Each question is worth a certain number of points. Generally, the long-response questions are worth 10 points, and the short-response questions are worth either 3 or 4 points each.

- Your responses must be written in paragraph form; outlines are not acceptable. Diagrams can be used to supplement your writing but will not receive credit on their own (unless otherwise specified).

- Remember to keep your response focused and use specific data and examples (where applicable) to justify your answers.

- Familiarize yourself with the key terms in this book as well as the Big Ideas for the AP Biology course. Understanding these will help you write clear, focused, and accurate responses.

# REFLECT

Respond to the following questions:

- Which topics discussed in this book do you hope to see on the free-response section?

- Which topics discussed in this book do you not hope to see on the free-response section?

- Can you define the Key Terms at the end of each chapter in order to use them appropriately in a free response?

- Which content chapters will you review in order to help you write about those topics on the free-response section?

- Will you seek further help, outside this book (such as a teacher, tutor, or AP Central), on any of the content in this chapter—and if so, on what content?

# Chapter 15
# Laboratory

All AP Biology courses have a laboratory component that gives students hands-on experience regarding some of the biology topics covered in class. Through these laboratory exercises you can learn the scientific method, lab techniques, and problem-solving skills. College Board suggests that you complete at least 8 of the following 13 lab experiences, in order to reinforce some of the most important topics in biology.

How do these labs relate for the AP Biology Exam? Approximately 10 percent of the questions on the multiple-choice section and one essay question will refer to certain aspects of these 13 lab exercises. ETS incorporates questions concerning the labs to determine two things: (1) how well you understand the key concepts in biology and (2) how well you think analytically. ETS wants to find out if you can design experiments, interpret data, and draw conclusions from these experiments.

There are basically two ways to approach a review of the laboratory exercises for this exam. We could review every single experiment, including the procedures and results. Not only would this take up more than half the book, it wouldn't necessarily improve your score.

We at The Princeton Review have a better approach. Because schools are not required to finish every single lab, we'd rather not overwhelm you with all the boring details about each and every experiment. Instead, we're going to tell you the objective of each lab. That way, even if you haven't completed all 13 labs, you will still be aware of the important concepts and results related to each experiment. How should you use this chapter to review for the AP Biology Exam? Read each summary and focus on the concepts stressed in each experiment. Now, let's get cracking!

# LAB 1: ARTIFICIAL SELECTION

This lab explores artificial selection—the process by which humans decide traits to enhance or diminish in other species by crossing individuals with the desired phenotype. The principles that you need to understand are the basic principles of natural selection and how natural selection drives evolution, which are:

- Variation is present in any population.

- Natural selection, or differential reproduction in a population, is a major mechanism in evolution.

- Natural selection acts on phenotypic variations in populations. Some organisms will have traits that are more favorable to the environment and will survive to reproduce more than other individuals, causing the genetic makeup of the population to change over time.

This lab specifically deals with Wisconsin Fast Plants, and in order to artificially select for certain traits, plants with the desired traits can be crossed, changing the genetic makeup of the population. For example, if you wanted to select for height, you could cross only the tallest plants with one another. The new population will have a higher mean height than the previous generation.

# LAB 2: MATHEMATICAL MODELING: HARDY-WEINBERG

In order to understand everything you need for the AP Biology Exam, you need to be able to use the Hardy-Weinberg principles and equations to determine allele frequencies in a population. One way to study evolution is to study how the frequencies of alleles change from generation to generation.

- Know how to calculate the allele and genotype frequencies using the two Hardy-Weinberg equations: $p + q = 1$ and $p^2 + 2pq + q^2 = 1$. Don't forget: If the population obeys Hardy Weinberg's rules, these frequencies remain constant over time.

- Review the discussion of the Hardy-Weinberg principle in this book. Know the five conditions of the Hardy-Weinberg equilibrium: (1) large population, (2) no mutations, (3) no immigration or emigration, (4) random mating, and (5) no natural selection.

- Review natural selection and how it can lead to changes in the genetic makeup of a population.

# LAB 3: COMPARING DNA SEQUENCES TO UNDERSTAND EVOLUTIONARY RELATIONSHIPS WITH BLAST

This laboratory uses BLAST (Basic Local Alignment Search Tool), a database that allows you to input a DNA sequence for a gene to look for similar or identical sequences present in other species. In order to understand everything you need for the AP Exam you need to be able to do the following:

- Be able to look at a phylogenic tree, which is a way to visually represent evolutionary relatedness. Endpoints of each branch correspond to a specific species, and each junction on the tree represents a common ancestor. Species that are closer on a phylogenic tree are more closely related.

- Understand BLAST scoring. Most phylogenic trees are constructed by examining nucleotide sequences; the more identical two species sequences are for a specific gene, the more closely related they are. BLAST is able to analyze different sequences to tell you how similar they are to one another based on a score. The higher the score, the closer the two sequences align.

# LAB 4: DIFFUSION AND OSMOSIS

This lab investigates the process of diffusion and osmosis in a semipermeable membrane as well as the effects of solute concentration and water potential on these processes.

What are the general concepts you really need to know?

Fortunately, this lab covers the same concepts about diffusion and osmosis that are discussed in this book. Just remember that osmosis is the movement of water across a semipermeable membrane from a region of high water concentration to one of low water concentration from a hypotonic region (low solute concentration) to a hypertonic region (high solute concentration).

- Be familiar with the concept of osmotic potential, which is simply the free energy of water. It is a measure of the tendency of water to diffuse across a membrane. Water moves across a selectively permeable membrane from an area of higher osmotic potential to an area of lower water potential.

- Be familiar with the effects of water gain in animal and plant cells. In animals, the direction of osmosis depends on the concentration of solutes both inside and outside of the cell. In plants, osmosis is also influenced by turgor pressure—the pressure that develops as water presses against a cell wall. If a plant cell loses water, the cell will shrink away from the cell wall and plasmolyze.

Another important concept to understand is the importance of surface area and volume in cells. Cells maintain homeostasis by regulating the movement of solutes across the cell membrane. Small cells have a large surface area to volume ratio, however as cells become larger, this ratio becomes smaller, giving the cell relatively less surface area to exchange solutes. A cell is limited in size by the surface area to volume ratio. There are many organisms that have evolved strategies for increasing surface area, like root hairs on plants and villi in the small intestines of animals.

# LAB 5: PHOTOSYNTHESIS

The chemical equation for photosynthesis is:

$$6CO_2 + 6H_2O \rightarrow C_6H_{12}O_6 + 6O_2$$

Because plants consume some of this energy during photosynthesis, measuring the oxygen produced by a plant can tell us about the net photosynthesis that is occurring. In this laboratory, photosynthesis rates are measured using leaf discs that begin to float as photosynthesis is carried out, allowing you to see that photosynthesis is occurring.

There are several properties that affect the rates of photosynthesis, including:

- light intensity, color, and direction
- temperature
- leaf color, size, type

Be able to hypothesize about the effects of these variables: For example, as light intensity increases, so does the rate of photosynthesis. Remember, plants and animals both contain mitochondria and carry out cell respiration!

# LAB 6: CELLULAR RESPIRATION

In this lab the respiratory rate of germinating and nongerminating seeds, as well as small insects, is investigated. The equation for cellular respiration is:

$$C_6H_{12}O_6 + O_2 \rightarrow 6CO_2 + 6H_2O$$

Germinating seeds respire and need to consume oxygen in order to continue to grow. Non-germinating seeds do not respire actively. In this lab, the amount of oxygen consumed by these types of seeds was measured using a respirometer. The experiment was also conducted at two temperatures, 25°C and 10°C, because seeds consume more oxygen at higher temperatures. You should know the following concepts:

- Oxygen is consumed in cellular respiration.
- Germinating seeds have a higher respiratory rate than non-germinating seeds, which have a very low respiratory rate.
- Know how to design a study to determine the effect of temperature on cell respiration.
- Be able to explain the significance of a control using glass beads. A control is a condition held constant during an experiment. In this case, glass beads are used as a control because they will not consume any oxygen.

# LAB 7: MITOSIS AND MEIOSIS

This lab highlights the differences between mitosis and meiosis. In this lab, slides of onion root tips were prepared to study plant mitosis. The important information and skills to review in this lab include the following:

- Mitosis produces two genetically identical cells, while meiosis produces haploid gametes.

- Cell division is highly regulated by checkpoints that depend, in part, on complexes of proteins called cyclins with other proteins called cyclin-dependent kinases. One such example is mitosis promoting factor (or MPF) that has its highest concentration during cell division and thought to usher a cell into mitosis.

- Nondisjunction, or the failure of chromosomes to separate correctly, can lead to an incorrect umber of chromosomes (too many or too few) in daughter cells.

- Know what each phase of the cell cycle looks like under a microscope (i.e., during metaphase, the chromosomes are aligned on the metaphase plate). Chapter 9 contains diagrams of each phase.

In one section of this lab, the sexual life cycle of the fungus Sordaria fimicola was examined. Sexual reproduction in this fungus involves the fusion of two nuclei: a (+) strain and a (–) strain—to form a diploid zygote. This zygote immediately undergoes meiosis to produce asci, which each contain eight haploid spores.

- During meiosis, crossing-over can occur to increase genetic variation. If crossing-over, or recombination has occurred, different genetic combinations will be observed in the offspring when compared with the parent strain.

- These offspring with new genetic combinations are called recombinants. By examining the numbers of recombinants with the total number of offspring, an estimate of the linkage map distance between two genes can be calculated using the following equation:

Map distance (in map units) = [(# recombinants) / (# total offspring)] × 100.

# LAB 8: BIOTECHNOLOGY: BACTERIAL TRANSFORMATION

In this lab, the principles of genetic engineering are studied. Biotechnologists are able to insert genes into an organism's DNA in order to introduce new traits or phenotypes, like inserting genes into a corn genome that help the crops ward off pests. This process is very complicated in higher plants and animals but relatively simple in bacteria. You are responsible for knowing the ways in which bacteria can accept fragments of foreign DNA.

- Conjugation: the transfer of genetic material between bacteria via a pilus, a bridge between the two cells

- Transformation: when bacteria take up foreign genetic material from the environment

- Transduction: when a bacteriophage (a virus that infects bacteria) transfers genetic material from one bacteria to another

In addition, DNA can also be inserted into bacteria using plasmids, which are small circular DNA fragments that can serve as a vector to incorporate genes into the host's chromosome. Plasmids are key elements in genetic engineering. The concepts you need to know about plasmids include the following:

- One way to incorporate specific genes into a plasmid is to use restriction enzymes, which cut foreign DNA at specific sites, producing DNA fragments. A specific fragment can be mixed together with a plasmid, and this recombinant plasmid can then be taken up by *E. coli*.

- Plasmids can give a transformed cell selective advantage. For example, if a plasmid carries genes that confer resistance to an antibiotic like ampicillin, it can transfer these genes to the bacteria. These bacteria are then said to be transformed. That means if ampicillin is in the culture, only transformed cells will grow. This is a clever way scientists can find out which bacteria have taken up a plasmid.

- In order to make a bacteria cell take up a plasmid, you must (1) add $CaCl_2$, (2) heat shock the cells, and (3) incubate them in order to allow the plasmid to cross the plasma membrane.

## LAB 9: BIOTECHNOLOGY: RESTRICTION ENZYME ANALYSIS OF DNA

This laboratory introduces you to the technique of electrophoresis. This technique is used in genetic engineering to separate and identify DNA fragments. You need to know the steps of this lab technique for the AP exam.

1.  DNA is cut with various restriction enzymes.

2.  The DNA fragments are loaded into wells on an agarose gel.

3.  As electricity runs through the gel, the fragments move according to their molecular weights. DNA is a negatively charged molecule; therefore, it will migrate toward the positive electrode. The longer the DNA fragment, the slower it moves through the gel.

4.  The distance of each fragment is recorded.

Restriction mapping allows scientists to distinguish between the DNA of different individuals. Since a restriction enzyme will only cut a specific DNA sequence, it will cause a unique set of fragments for each individual called restriction fragments length polymorphisms, or RFLPs for short. This technology is used at crime scenes to help match DNA samples to suspects.

## LAB 10: ENERGY DYNAMICS

This lab examines energy storage and transfer in ecosystems. Almost all organisms receive energy from the sun either directly or indirectly.

*   Producers, or autotrophs, are organisms that can make their own food using energy captured from the sun. They convert this into chemical energy that is stored in high-energy molecules like glucose.

*   Consumers, or heterotrophs, must obtain their energy from organic molecules in their environment. They can then take this energy to make the organic molecules they need to survive.

*   Biomass is measure of the mass of living matter in an environment. It can be used to estimate the energy present in an environment.

# LAB 11: TRANSPIRATION

This lab investigates the mechanisms of transpiration, the movement of water from a plant to the atmosphere through evaporation. What do you need to take away from this lab?

- The special properties of water that allow it to move through a plant from the roots to the leaves include polarity, hydrogen bonding, cohesion, and adhesion

- The vascular tissues that are involved in transport in plants. Xylem transport water from roots to leaves, while phloem transport sugars made by photosynthesis in the leaves down to the stem and roots.

- Stomata are small pores present in leaves that allow $CO_2$ to enter for photosynthesis and also are a major place where water can exit a plant during transpiration.

# LAB 12: FRUIT FLY BEHAVIOR

In this lab, fruit flies are given the choice between two environments using a choice chamber, which allows fruit flies to move freely between the two. Typically, fruit flies prefer an environment that provides either food or a place to reproduce. They also respond to light and gravity.

- Taxis is the innate movement of an organism based on some sort of stimulus. Movement toward a stimulus is called positive taxis, while movement away from a stimulus is negative taxis. In this lab, fruit flies exhibited a negative gravitaxis (or a movement opposite to the force of gravity) and positive phototaxis (a movement toward a light source).

# LAB 13: ENZYME ACTIVITY

This lab demonstrated how an enzyme catalyzes a reaction, and what can influence rates of catalysis. In this lab the enzyme peroxidase was used to catalyze the conversion of hydrogen peroxide to water and oxygen.

$$H_2O_2 + \text{peroxidase} \rightarrow 2H_2O + O_2 + \text{peroxidase}$$

The following are the major concepts you need to understand for the AP exam:

- Enzymes are proteins that increase the rate of biological reactions. They accomplish this by lowering the activation energy of the reaction.

- Enzymes have active sites, which are pockets that the substrates (reactants) can enter that are specific to one substrate or set of substrates.

- Enzymes have optimal temperature and pH ranges at which they catalyze reactions. Enzyme concentration and substrate concentration can also influence the rates of reaction.

- If you're asked to design an experiment to measure the effect of these four variables on enzyme activity, keep all the conditions constant except for the variable of interest. For example, to measure the effects of pH in an experiment, maintain the temperature, enzyme concentration, and substrate concentration as you change pH.

# REFLECT

Respond to the following questions:

- For what laboratory concepts discussed in this chapter do you feel you have achieved sufficient mastery to answer multiple-choice questions correctly?

- For which laboratory concepts discussed in this chapter do you feel you have achieved sufficient mastery to discuss effectively in a free response?

- For which laboratory concepts discussed in this chapter do you feel you need more work before you can answer multiple-choice questions correctly?

- For which laboratory concepts discussed in this chapter do you feel you need more work before you can discuss effectively in a free response?

- What parts of this chapter are you going to re-review?

- Will you seek further help, outside this book (such as a teacher, tutor, or AP Central), on any of the content in this chapter—and if so, on what content?

# Chapter 16
# Chapter Drill
# Answers and
# Explanations

# CHAPTER 4 DRILL: THE CHEMISTRY OF LIFE

1. **C** Water has many unique properties that favor life, including (A), a high specific heat, (B), high surface tension and cohesive properties, and (D), high intermolecular forces due to hydrogen bonding. However, water is a very polar molecule and is an excellent solvent, making (C) inaccurate and the correct answer.

2. **B** The pH scale is logarithmically based, meaning that each difference of 1 on the log scale is indicative of a ten-fold difference in the hydrogen ($H^+$) ion concentration. According to the question, the pH values of the cytoplasm of cells and gastric juices are approximately 7.4 and 2.5, respectively. Therefore, the pH values vary by 5 and the hydrogen ($H^+$) ion concentrations much vary by 5 orders of ten ($10^5$) or 100,000-fold. Since lower pH values have more hydrogen ions, they are also more acidic.

3. **C** All amino acids share a carboxylic acid group, COOH, labeled (B), and an amino-group, $NH_2$, labeled (D). They also share a bound hydrogen atom to the central carbon (A). Differences in amino acids are defined by variations in the fourth position called the R group, labeled (C).

4. **B** The hypothesis that life may have arisen from formation of complex molecules from the primordial "soup" of Earth is not supported by the absence of nucleic acids. All life is DNA-based, yet no nucleic acid molecules were detected. The presence of carbon molecules, amino acids, and sugars, which are common compounds and comprise life, supports the hypothesis, so eliminate (A), (C), and (D). Choice (B) is correct.

5. **C** The amino acids cysteine and methionine contain the element sulfur (as indicated by the S in the amino acid structure shown). However, no sulfur-based compounds were included in the Miller-Urey experiment, so it was impossible to form these two amino acids under the conditions of the experiment.

6. **B** Silica is a mineral form of glass, is not a common component of life forms, and is largely chemically inert. Since oxygen is already present in several compounds included in the experiment, addition of this compound does not provide any additional elements or chemical substrates, which would permit generation of additional amino acids or synthesis of nucleic acids. The addition of sulfur compounds and phosphorus are necessary to generate some amino acids and all nucleic acids, which eliminates (A), (C), and (D).

7. **D** Hydrolysis adds water to a polymer to break the linkages between monomers. Therefore, (A), free water, will not result because water is being broken down in the reaction. Free water would result from the opposite reaction, for condensation reactions. Choice (B), adenine, results from the hydrolysis of DNA, as it is a monomer of DNA. Choice (C), cholesterol, is a steroid and therefore does not undergo hydrolysis. Dipeptides could result from the hydrolysis of proteins, which are composed of polypeptides.

8.  **B**  Phospholipids have a long fatty acid tail and a polar head group. The fatty acid tails associate in the membrane, and the phospholipid head group associates with water at the boundary of the membrane. Choice (A), nucleotides, are mostly polar and found in DNA and RNA. Choice (C), water, is polar and not found in plasma membranes. Choice (D), amino acids, are zwitterionic, with both a positive and a negative charge. Free amino acids are not found in the membrane.

9.  **C**  $(CH_2O)_3 = C_{18}H_{36}O_{18}$ (A), but the dehydration reactions to produce two glycosidic bonds between the three monosaccharides would remove two $H_2O$ molecules, or four hydrogens and two oxygens. Therefore, the correct answer is (C).

10. **A**  All isotopes of hydrogen will contain one proton, by definition. All atoms that only have one proton are identified as hydrogen. Tritium is an isotope that has one proton and two neutrons, for a total of three particles in its nucleus. Choice (A) is correct. Choice (B) is not correct because two protons would mean that the element is helium, not hydrogen. Choice (C) is not correct because the atomic number (1) will never change without changing the identity of the atom. Finally, (D) is not correct because radioactive atoms do not give off electrons.

# CHAPTER 5 DRILL: CELLS

1.  **C**  Active transport is the movement of substances across a membrane against their concentration gradient through the use of energy (ATP). The sodium-potassium pump is a critical structure which uses ATP hydrolysis to move sodium and potassium ions against their respective concentration gradients. Diffusion of oxygen is an example of simple diffusion and can occur without the need of channels or pores. Water uses aquaporins to travel across the membrane, down its concentration gradient; therefore, it also does not require energy. Movement of sodium ions by a voltage-gated ion channel (B) is an example of facilitated diffusion. Because the sodium ions are still undergoing diffusion from high to low concentrations without the need of ATP, this does not represent a form of active transport.

2.  **C**  Bacterial cells can be visualized using light microscopy. In fact, back in the seventeenth century, some of the earliest studies using primitive microscopes recorded the shape and organization of bacteria. Choices (A), (B), and (D) are incorrect, as viruses and cell organelle structure are too small to be observed using light microscopy and require electron microscopy.

3.  **D**  The cell wall is a structure that is present in bacteria but absent in animal cells. Consequently, this structure is targeted by several leading classes of antibiotics and would be an effective target of therapeutics against *V. cholerae*. Flagella, plasma membrane, and ribosomes—(A), (B), and (C), respectively—are all structures that are present in animal cells.

4.  **C**   Based on the microscopy data, the organism has a cell wall and lacks mitochondria. The presence of a cell wall suggests that the organism is not a protozoan, so eliminate (B). The absence of mitochondria is indicative of prokaryotic structure because they lack organelles, eliminating (A) and (D). The most likely conclusion is that this organism is a bacterial species.

5.  **700**   The graphed line of chlorophyll a has its highest peak at 700 nm.

6.  **B**   Although many of the mentioned organelles work closely together, the only choice that can be correct is (B). Ribosomes translate and manufacture proteins, often on the rough endoplasmic reticulum. Those proteins are then packaged in the Golgi bodies to vesicles, which are then fused with the plasma membrane. Choice (A) is incorrect because the nuclear envelope and nucleolus do not interact with vacuoles; they interact with the centrioles only during mitosis. Choice (C) is incorrect because ribosomes don't often interact with ribosomes, and lysosomes only interact with mitochondria and chloroplasts when they degrade those organelles. Choice (D) is incorrect because the nucleolus doesn't interact with the smooth endoplasmic reticulum, lysosomes, or centrioles.

7.  **C**   A noncompetitive inhibitor would bind to an estrogen receptor, reducing the effectiveness of this binding. Choice (A) is incorrect because estrogen binds to intracellular receptors, not receptors at the plasma membrane. Choice (B) is incorrect because testosterone and estrogen have different effects, but testosterone would not change estrogen effectiveness. Choice (D) is also incorrect because estrogen is not secreted from the pituitary gland; therefore, it would not have an effect on estrogen levels.

8.  **B**   You would need a cell that makes and secretes a lot of protein or hormones. Choice (C), *E. coli*, do not have organelles. Choice (D), leukocytes, is not the best answer because leukocytes do not secrete large amounts of hormones or proteins. Choice (A), neurons, is a potential answer, but (B) is the best answer because the pancreas is a gland that secretes copious amounts of protein and hormones.

9.  **A**   Choice (A) is correct because contractile vacuoles expel water that accumulates in cells in a hypotonic environment. The strategy described in (B) is wrong because it would increase water intake in *Paramecium*. Choice (C) is true for *Paramecium* but does not help with water gain. Choice (D) could help *Paramecium* find less challenging environments, but then it could be in danger of drying out.

10. **C**   ATP is consumed by the $Na^+K^+$ sodium potassium pump, so (A) is incorrect. Choice (B) is incorrect because, again, ATP is consumed. Choice (D) is incorrect because cotransporters can go in one direction or two directions. Choice (C) is the correct answer because ATP drives transport in the $Na^+K^+$ sodium potassium pump.

# CHAPTER 6 DRILL: CELLULAR ENERGETICS

1. **B** The Krebs cycle primarily occurs in the matrix of the mitochondria. The inner membrane and the intermembrane space—(A) and (C), respectively—are used in oxidative phosphorylation.

2. **D** Inhibitor Y is binding at a site outside of the active site and inducing a conformational change in the enzyme structure. By binding outside of the active site, it must be an allosteric inhibitor, eliminating (A) and (C). Because the inhibitor is binding outside of the active site, it is not competing with the substrate for binding, so it is considered a non-competitive inhibitor.

3. **B** Enzymes are biological catalysts, which lower the activation energy (the energy threshold that must be met to proceed from reactant to product). The reaction coordinate diagram must reflect a decrease in the activation energy, eliminating (C). Furthermore, the enzyme does not alter the energy of the reactants or products, eliminating (A) and (D).

4. **B** Based on the pathway provided, consumption of 1 glucose and 2 ATP results in production of 4 ATP. In other words, each glucose results in a net gain of 2 ATP. Therefore, 2 glucose molecules would result in a net gain of 4 ATP.

5. **D** $NAD^+$ is a required cofactor for glycolysis and must be regenerated to permit glycolysis to continue to occur. In fermentation, pyruvic acid is converted into either ethanol or lactic acid. During this process, NADH is recycled into $NAD^+$.

6. **B** Glycolysis results in the production of ATP (energy), so it is considered an exergonic process.

7. **B** Thermophilic bacteria live in hot environments. Also, a DNA polymerase replicates DNA. Therefore, in the PCR technique, the stage in which DNA is elongated by a DNA polymerase is part 3 of the cycle, at 72°C. Therefore, bacteria growing in hydrothermal vents between 70–75°C is most likely the answer. The conditions described in (A) and (D) do not reflect the data, which focuses on temperature. Choice (C) describes hot springs, but the temperature does not reflect the enzyme activity most likely for *Taq* polymerase.

8. **D** The free energy does not change between catalyzed and uncatalyzed reactions; therefore, (A) is incorrect. Choice (B) is a correct statement, but it does not answer the question. Choice (C) is incorrect because enzymes catalyze reactions.

9. **B** ATP production will increase in the treated mitochondria because the low pH provides more $H^+$ ions in the solution. Also, oxygen provides a terminal electron acceptor for oxidative phosphorylation.

10. **B** Oxygen is consumed at the end of the electron transport chain, where electrons are combined with water to transport $H^+$ and liberate oxygen. Oxygen consumption, therefore, means oxidative phosphorylation is occurring in the mitochondria. Reduced oxygen consumption indicates that the inhibitor is working; therefore, (A) and (D) are incorrect. Choice (C) is unlikely because oxygen is consumed in the mitochondria. Choice (B) is correct because $FADH_2$ donates electrons to the complex II, bypassing the inhibition of rotenone in complex I. Only NADH electrons will be blocked.

11.  **D**  Choices (A) and (B) are clearly wrong, as they defy physics. Choice (C) is true but unrelated. Choice (D) is the correct answer because it explains the increase in entropy. Even though organisms build and develop as ordered systems, heat is lost continuously. Additionally, organisms exhale gases and produce waste products that balance the effect of order.

# CHAPTER 7 DRILL: PHOTOSYNTHESIS

1.  **C**  Antenna pigments capture light and transfer its energy to the reaction centers where photosystems I and II exist.

2.  **D**  The light-independent reactions take $CO_2$ and NADPH to generate glucose. As a byproduct of this reaction, $NADP^+$ is generated, which must be recycled into NADPH by the light-dependent reactions. $NAD^+$ is associated with mitochondrial activity, so eliminate (B). Choices (A) and (C) are also incorrect because NADPH and ATP are generated by light-dependent reactions.

3.  **C**  Carbon dioxide is used during the light-dependent reactions referred to as the Calvin cycle or $C_3$ pathway to generate glucose.

4.  **B**  Noncyclic photophosphorylation results in production of NADPH, which is subsequently used in the light-independent reactions to generate glucose. Cyclic photophosphorylation only produces ATP, not NADPH, and therefore is less efficient as a photophosphorylation process.

5.  **D**  Under conditions where ferredoxin is inhibited, photosystem II will still be active in the noncyclic photophosphorylation cascade. A byproduct of photosystem II is the production of free $O_2$ from the split of water using sunlight.

6.  **A**  A key fundamental difference between cyclic and noncyclic photophosphorylation is the photolysis of water. In cyclic photophosphorylation, water is not split and no NADPH is produced. However, in noncyclic photophosphorylation, photolysis occurs and water is split to release free $O_2$ gas.

7.  **B**  Choice (A) is incorrect because ATP is generated in the stroma, on the outside of the thylakoid lumen. Choice (C) is incorrect. NADPH is produced at the termination of the electron transport chains in the thylakoid lumen. Choice (D) is incorrect because chloroplasts do not consume oxygen as mitochondria do.

8.  **D**  Choice (A) is incorrect because protons are pumped across the thylakoid membrane to generate ATP, not NADPH. Choice (B) is incorrect because ATP synthesis occurs in the matrix and the stroma, not the lumen. Choice (C) is incorrect because protons are pumped against their concentration gradient in both chloroplasts and mitochondria.

9.  **B**  In sunlight, oxygen will be produced as photosynthesis occurs. In green light, the cells will not absorb as much light because green light is reflected, not absorbed by chlorophyll.

10.  **C**  This $C_3$ plant does not have any strategies to handle those conditions, so (C) is the answer. $C_4$ and CAM plants are designed to use PEP carboxylase to avoid photorespiration.

# CHAPTER 8 DRILL: MOLECULAR GENETICS

1.  **B**  Okazaki fragments are generated during DNA replication when the DNA polymerase must create short DNA segments due to its requirement for 5' to 3' polymerization. Since the newly discovered yeast cell has 3' to 5' activity, there would be no lagging strand and likely no Okazaki fragments.

2.  **C**  Since the gene is much shorter than expected, a stop codon must have been introduced by mutagenesis. This is an example of a nonsense mutation.

3.  **D**  The order for DNA replication is helicase, RNA primase, DNA polymerase, and ligase.

4.  **C**  If an mRNA codon is UAC, the complementary segment on a tRNA anticodon is AUG.

5.  **D**  During post-translational modification, the polypeptide undergoes a conformational change. Choices (A), intron excision, and (B), poly(A) addition, are examples of post-transcriptional modifications, and formation of peptide bonds occurs during translation, not afterwards.

6.  **B**  If 21 nucleotides comprise a sequence and 3 nucleotides comprise each codon, there would be 7 codons and thus a maximum of 7 amino acids.

7.  **A**  Choice (B) is incorrect because bacteria make membrane proteins that reside on the plasma membrane. Choice (C) is also incorrect because bacteria use transcription factors. Choice (D) is possible, but (A) is the best answer.

8.  **C**  Choice (A) is likely incorrect. Choice (B) would be deleterious for the cell. Choice (D) is also incorrect, as DNA replication occurs only during or preceding binary fission.

9.  **B**  Viruses do not have their own ribosomes, flagellum, or independent metabolism, so (A), (C), and (D) are incorrect. The only possible answer is (B).

10. **C**  Transformation occurs when bacteria take up DNA from their surroundings. The pathogenic bacteria did not come alive again, as suggested by (A). Protein was not the transforming agent, so (B) is incorrect. Finally, (D) does not make sense, as genes cannot turn into other genes simply by being in a different cell context.

11. **C**  Choice (C) describes the result of semiconservative replication, which is correct. In semiconservative replication, each heavy strand will combine with a light strand. Therefore, all of the DNA will be half-heavy and half-light, resulting in a band with a density averaging the two densities. Choices (A), (B), and (D) describe the result of conservative replication, which does not occur.

12. **B**   If there are DNA/RNA fragments, then helicase (A) is present and has opened the strands. RNA primase (D) is also present because RNA primers have been added. DNA polymerase (B) is missing because it has not elongated the strand. DNA ligase (C) could be missing, but it is impossible to tell because DNA has not been elongated by DNA polymerase. Therefore, (B) is the best answer.

# CHAPTER 9 DRILL: CELL REPRODUCTION

1. **C**   During anaphase, the chromatids are separated by shortening of the spindle fibers. Chemically blocking the shortening of these fibers would arrest the cell in metaphase. The cells are arrested in metaphase as indicated by the alignment of the chromosomes in the center of the cell and their attachment to spindle fibers, eliminating (A) and (D). The next phase (anaphase) involves separation of the chromatids during shortening of the spindle fibers. Therefore, the cells are not arrested due to dissociation of the kinetochore from the spindle fibers, eliminating (B).

2. **A**   Crossing-over occurs during prophase I and constitutes the key step which creates exchange of alleles during meiosis.

3. **C**   Lactose is facilitating transcription by removing a molecule which blocks RNA polymerase activity; therefore it is an inducer. The molecular target of lactose is a repressor, which prevents the RNA polymerase from reading through a gene and synthesizing RNA.

4. **C**   The synthesis, or S phase, of the cell cycle represents the step in which the genetic material is duplicated. The only phase labeled in the experiment that represents an increase is phase B. Based on the time scale on the *x*-axis, this phase lasts approximately 30 min.

5. **D**   Anaphase represents the cell division stage of the cell cycle and would be the phase that occurs right before the amount of genetic material should decrease. Phase D is the phase right before the genetic material would drop, so (D) is the correct answer.

6. **52**   The organism has 13 chromosomes in gametic cells, which are haploid. Therefore, a diploid cell would normally have two copies of each chromosome or 26 chromosomes. During S phase, the cell must make a duplicate copy of all of its chromosomes, so the cell would contain 52 chromosomes.

7. **D**   Choices (A) and (B) do not occur, but if they did, they would not impact the gametes. Choice (C) is not likely to occur because the mitotic spindles attach only to kinetochore protein complexes at the centromere. Therefore, (D) is the answer.

8. **C**   Metaphase would stall because the chromosomes would be stuck at the metaphase plate, waiting for the mitotic spindle to attach. Therefore, (C) is correct. Choice (A) is incorrect because the mitotic spindle is not involved at that stage of the cell cycle. Choices (B) and (D) are incorrect, as the mitotic spindle would be required before these phases occurred.

9.  **B**  Both (A) and (B) have sperm or ovum as the first type of cell. Because these are both gametes, they will both have half of the DNA of other types of cells. Choices (C) and (D) can thus be eliminated. Muscles and neurons are both terminally differentiated in $G_0$ arrest. However, liver and taste buds potentially differ. Although liver cells can divide, taste buds divide at a higher rate, so they are most likely to be in $G_2$ phase. Therefore, the answer is (B).

10.  **A**  Nondisjunction results in major changes to the genome, so (B) and (C) can be eliminated. Deletion of an enhancer, (D), could affect the gene expression and then phenotype. However, in (A), translocation may not produce any effect, as the same information exists in the genome.

# CHAPTER 10 DRILL: HEREDITY

1.  **D**  The father and mother are both AB blood type. Since neither parent has a recessive allele, it is impossible for their child to be O blood type.

2.  **B**  When the phenotype associated with two traits is mixed, this is considered an example of incomplete dominance. In this case, neither red nor blue color was dominant, and the resulting progeny exhibited a mixture of the traits (purple).

3.  **B**  Crossing the pea plant that is heterozygous for both traits (TtGg) with a plant that is recessive for both traits (ttgg) results in the following possible combinations, each of which should occur 25% of the time: TtGg (Tall and Green), Ttgg (Tall and Yellow), ttGg (Short and Green), and ttgg (Short and Yellow).

4.  **C**  Because the woman is a carrier, she must have one normal copy of the X chromosome and one diseased copy. Since the boy will receive an X chromosome from his mother, there is a 50% chance that he will receive a diseased copy. Because he doesn't have a second X chromosome, he must have the disease if he receives the diseased X chromosome.

5.  **D**  They both must have a normal copy of the X chromosome. It is possible that the woman may be affected with hemophilia; however, that scenario is extremely unlikely because such a case would require two diseased copies of the X chromosome.

6.  **D**  Essentially, transmission of hemophilia to a girl born with Turner Syndrome would be very similar to the conditions by which a boy would receive the disease. Both boys and girls with Turner Syndrome only have one copy of the X chromosome. Therefore, they both would have a 50% chance of receiving the diseased copy of the X chromosome.

7.  **B**  Choice (A) is true, but it doesn't explain why the parents are normal. Choice (B) is a better explanation. Carriers of recessive genetic diseases frequently do not have a recognizable phenotype because one normal allele provides enough functioning protein to avoid the ill effects of the disease allele. Choice (C) is also possible, but (B) is still the better answer. Finally, (D) is not correct because CLN3 is an autosomal gene.

8.  **C**  Choice (C) describes a testcross that will actually let the breeder know if the male is heterozygous. If any white spotted pups result from that cross, then the male would have contributed to a spotted allele. Choice (B) describes a male that is homozygous for the spot allele, so it's incorrect. Choice (A), stop breeding Speckle, is a way to remove spots from the line, but it will not help you identify males with the spot allele. Choice (D) also will not help you determine which black Labrador males have a spot allele.

9.  **A**  Choice (B) does not apply to this question. Choice (C) refers to a gene not following Mendel's law of dominance. Choice (D) is a reason why traits would not follow independent assortment, but nondisjunction would cause many traits—whole chromosomes—to differ from the law. Choice (A)—genes are linked on the same chromosome—describes a normal reason for why certain traits would not follow Mendel's law of independent assortment.

10. **A**  Choices (B) and (C) are unlikely to be true and can be eliminated. Choice (D) is true but does not answer the question. Choice (A) is the easiest and most probable explanation.

# CHAPTER 11 DRILLS

## Lymphatic and Immune Systems Drill

1.  **B**  Lymph nodes harbor large quantities of lymphocytes. During active infections or cancer, mass expansion of lymphocytes causes enlargement and swelling of lymph nodes. This is why doctors often feel the areas around lymph nodes nearest to where a suspected infection is.

2.  **C**  The placebo strain was included as a control for the experiment. Because no active antibodies should be produced to the placebo strain, the scientists would be able to compare antibodies elicited specifically to the vaccine strain.

3.  **B**  B-cells (or plasma cells) are responsible for creating antibodies. These are the primary cells associated with lasting memory to vaccinations.

4.  **A**  The most practical way to increase statistical significance is to increase the sample size, which is (A). More control variables will only ensure that they haven't introduced bias. Repeating a third challenge would evaluate the length of potential protection, not the extent. Using a different statistical test would simply manipulate the data and not provide true significance.

5.　**C**　Choice (A) could work, but it would cause the entire immune system to be reduced, which could be harmful to the patient and is not a targeted response. Choice (B), removal of the thymus, is a treatment for this disease (the thymus is where the maturation of T-cells take place), but it is incorrect because macrophages do not mature in the thymus. Choice (D) is incorrect because additional acetylcholine will not make a difference because the receptors are affected. Choice (C) is the best answer because removing acetylcholine autoantibodies could reduce the damage to the receptors.

6.　**D**　Natural T-killer cells are the cells that damage a target cell, so (D) is correct. B-cells (A) and lymphocytes (B) are less specialized for destruction. Phagocytes (C) is a general term and not as good an answer as (D).

7.　**B**　Thorns and spines (B) are examples of innate defenses that plants have to avoid herbivores. Choice (C), attractive scents, would have the opposite effect. Stomata (D) allow gas exchange, which are unrelated to the question. Long stems (A) may be considered a defense only against certain types of threats. Therefore, (B) is the best answer.

8.　**A**　Choice (B) is incorrect because helper T-cells do not destroy pathogens. Choice (D) is incorrect, as the goal of the immune system is to prevent further infection. Choice (C) is incorrect because only memory B-cells remain, not plasma B-cells. Choice (A) is the correct answer because B-cells secrete antibodies specific to the pathogen in the complement system. This flags the pathogen for destruction.

9.　**B**　Choice (B) is the only logical answer, as the other three choices refer to different types of immune cells.

# Nervous System Drill

1.　**B**　The membrane potential depolarizes at −50 mV so that the voltage-gated sodium channels can open. Note that the sodium channels open and close more quickly than the potassium channels.

2.　**C**　The blocking of voltage-gated sodium ($Na^+$) channels by tetrodotoxin would prevent an action potential in cells and therefore the transmission of a neural response to the subsequent neurons.

3.　**A**　Upon seeing the bear charging towards her, Jane's sympathetic "fight or flight" response would be activated. Since this pathway is not voluntary, it is considered part of her autonomic nervous system.

4.　**B**　The cerebellum is responsible for coordinating the complex balance and motions of the body. Because all of the patient's muscles appear to be functionally active, the impairment would likely be limited to coordination rather than physical inhibition.

5.　**D**　Choices (A) and (C) describe situations that are not directly related to ouabain. Choice (B) is incorrect, as the channels that open during action potentials are separate from the sodium-potassium pumps. Choice (D) is correct.

6.   **B**   Proteins are made primarily in the cell body (B) of neurons. Dendrites (A) receive post-synaptic signals. Axons (C) and axon terminals (D) transmit action potentials in the pre-synaptic neuron. Choice (B) is correct.

7.   **C**   In a neuromuscular junction, the neuron synapses with a muscle. The neuron releases acetylcholine as a neurotransmitter to tell the muscle to contract. Thus, acetylcholinesterase removes extra acetylcholine from the synapse after the transmission. Without this enzyme, the neurotransmitter will build up in the synapse, causing overstimulation of the acetylcholine receptors on the muscle and hyper-contraction of the muscles. Choices (A), (B), and (D) are incorrect because they describe the opposite of what would occur.

8.   **A**   By definition, the threshold is the point at which enough voltage-gated channels have opened to permit a switch in the polarity of the membrane and an action potential to perpetuate through the axon of the neuron. Choices (B), the release of neurotransmitters, and (C), peak depolarization, both occur after the threshold is met. As in (D), sodium ions enter the cell during depolarization, but this choice is vague and not as strong as (A).

9.   **D**   GABA is the major inhibitory neurotransmitter in the central nervous system, so (D) is correct.

10.  **B**   Both Schwann cells and oligodendrocytes act to wrap around the axons of neurons to insulate the membrane depolarization and expedite the action potential. Choice (B) is correct.

11.  **C**   The cerebellum is responsible for motor coordination, which is lost in Batten's disease patients. The other answer choices reflect other regions of the brain, so they are incorrect.

## Endocrine System Drill

1.   **D**   Increased levels of vasopressin would not likely be observed since it is secreted by the posterior pituitary. Choices (A), (B), and (C) would all be triggered by hormones produced by the anterior pituitary.

2.   **D**   Calcitonin and parathyroid hormone (PTH) have antagonistic effects on blood calcium concentrations. Calcitonin lowers blood calcium levels, while PTH increases blood calcium levels.

3.   **B**   Thyroid stimulating hormone (TSH) is produced by the anterior pituitary gland to increase production of thyroid hormones.

4.   **D**   The hypothalamus is the part of the brain that regulates the pituitary gland and produces neurosecretory hormones. TSH and LH are hormones secreted by the anterior pituitary and the hypothalamus is not an extension of the pituitary gland.

5.  **B**  Choices (A), (C), and (D) are incorrect because testosterone and estradiol will easily diffuse across the membrane, as they are hydrophobic, steroid-type molecules. Receptors are not needed on the outside of the plasma membrane. Instead, receptors inside the cell will receive and transduce the signal. Choice (B) is the correct answer.

6.  **C**  Choice (C) is correct because insulin is secreted from the cells in the pancreas gland. Diabetes, which results from inadequate response to insulin, could result from improper function of pancreas cells, as is seen in type I and juvenile diabetes. Choice (A) is incorrect because muscular dystrophy is due to dysfunctional muscle and nerve cells. Choice (B) is also incorrect because gigantism is due to defects in growth hormone levels, which are derived from the pituitary gland. Finally, (D) is incorrect because dwarfism is also controlled by growth hormone, not hormones from the parathyroid gland.

7.  **A**  Glucagon is secreted when blood glucose levels are too low, so (A) is the answer. Therefore, a marathon runner usually has high levels of glucagon to liberate glucose from stores of glycogen in the body. Gastrin and insulin are secreted only after a meal, which is not likely in this scenario. Cortisol levels are high in stress situations, which is possible for a marathon runner; however, a conditioned runner would likely not have high levels of cortisol at the end of a race. The best response is (A).

8.  **B**  Choice (B) is the correct answer because the receptor for vasopressin is in the G-protein-coupled receptor family. These receptors sit on the plasma membrane. Only cells with the vasopressin receptor will respond to the presence of vasopressin, even if other cells are exposed to the hormone. Therefore, (A) and (C) are incorrect. Choice (D) is also incorrect because vasopressin does not enter cells.

9.  **D**  Of the answer choices, (D), insulin and glucagon, is the only pair with opposing effects. Insulin decreases the blood glucose levels, and glucagon increases the blood glucose levels. The other pairs do not have opposing effects on the body's homeostasis. Therefore, (D) is your answer.

10. **C**  Choice (A) is not correct because blood insulin levels are high in type II diabetes, but the cells are unresponsive—so the pancreas produces plenty of insulin. Choice (B) is not correct, as insulin receptors do not bind glucose directly; they bind insulin. Choice (D) is also incorrect because if glucagon levels were high, then blood sugar levels would rise even higher. Choice (C) is the best answer because repeated overstimulation of insulin receptors may cause them to be removed from the surface of cells.

## Reproductive System and Embryonic Development Drill

1. **B** The ectoderm will ultimately become the epithelial tissues and neural components of the organism. The GI tract will be associated with the endoderm, and the heart and pancreas will be associated with the mesoderm.

2. **B** Following ovulation, the corpus luteum is responsible for secreting progesterone to protect and enhance the endometrium for receiving a fertilized cell.

3. **C** The secretions of the prostate gland are alkaline and neutralize the acidic conditions of the vagina and uterus to prevent rapid death of sperm in the low pH conditions.

4. **A** Choice (B) is not true, as embryos do not "breathe" in this sense. Choice (C) is also incorrect because we did not evolve from birds. (However, humans and chickens, as well as other bird types, have a common ancestry, so their development is similar.) Choice (D) is incorrect because the gill slit structure does not develop into the respiratory system. Choice (A) is the correct answer.

5. **C** Choice (C) is correct. Neurons die by apoptosis if they do not find other neurons to connect with. Choice (A) is correct but unrelated to the question. Choice (B) describes a process that requires apoptosis but refers only to the final result, not the process. Also, (B) does not refer to roundworms. Choice (D) is incorrect because some cells that do not survive to adulthood had important functions in the developing organism. Once their stage of development passes, they are no longer necessary, and they die by apoptosis.

# CHAPTER 12 DRILL: EVOLUTION

1. **D** Mammals and cephalopods developed similar eye structures independently due to similar selective pressures. This is an example of convergent evolution.

2. **B** The data provided show a transition toward one extreme (black) and against another (white). This is an example of directional selection.

3. **B** Based on the data, the amount of white-bodied pepper moths decreased between 1802 and 1902, and the amount of black-bodied pepper moths increased during the same period.

4. **B** Longtail moths were included in the experiment as a control to compare the effects that are not associated with color.

5. **B** If the color of ash or soot produced by the Industrial Revolution were white or light gray, this would likely reverse the trend observed, applying additional selection against the black moths.

6.  **0.43** The frequency can be calculated as shown below:

$$p + q = 1$$
$$p + (0.3) = 1$$
$$p = 0.7$$
$$2pq = \text{Heterozygotes} = 2(0.7)(0.3) = 0.43$$

7.  **B** The mole-rats live in the same habitat, but over time there has been speciation, as the two groups have chosen different plants as their diet. Therefore, (B) is correct. Choice (A) is incorrect because there is not a habitat barrier; the mole-rats have access to each other. Choice (C) is incorrect because the mole-rats are capable of breeding, they just do not. Choice (D) is incorrect because the mole-rats do not interbreed.

8.  **C** Choice (A) is incorrect because this is a small population, so genetic drift is likely. Choice (B) is incorrect because this situation does not describe convergent evolution, and the second part of the answer choice does not accurately describe the change of the population. Choice (D) is incorrect because the number of red-haired people will likely increase, not decrease.

9.  **D** Choice (D) is the answer because it is the only choice that does not predict evolution occurring on the island. Choices (A), (B), and (C) apply to this situation, and they all predict evolution. The population is small, mutations are inevitable, and humans usually do not mate randomly. This population will definitely evolve or undergo genetic drift.

10. **A** Choice (A) is correct because the trait is being selected based on mating preferences of the females. Longer, bigger tails indicate reproductive fitness. Peahens choose males who are healthy and strong enough to grow these big tails and will hopefully produce the best offspring. Choices (B), (C), and (D) are incorrect because they do not describe the mating preferences of the females.

# CHAPTER 13 DRILL: BEHAVIOR AND ECOLOGY

1.  **D** The behavior displayed by the chimpanzee represents insight because the chimpanzee has figured out how to solve the problem without external influence or learning.

2.  **A** Viruses would display reproductive strategies most similar to r-strategists because they aim to reproduce as fast as possible and create as many progeny as possible, in order to increase their odds of transmission to other hosts.

3.  **C** They would be considered secondary consumers because they consume the primary consumers (plankton) which consume algae (the producers).

4.   **C**   Since brook trout can tolerate pH values as low as 4.9 and do not appear to diminish, the pH of the creek must exceed 4.9, so we can eliminate (D). Since crayfish cannot tolerate pH levels lower than 5.4, the pH of the stream must have dropped lower than this value. Only (C) falls in this range.

5.   **C**   Only acid rain would directly explain why the pH would drop in the creek over the time period.

6.   **D**   Based on Table 1, the pH must exceed 6.1 for snails to be able to return to the creek ecosystem.

7.   **D**   Choices (A) and (B) refer to light and gravity responses of plants. Choice (C) is not an actual tropism—it is a made-up word. Thigmotropism is the term for how plants respond to touch. The correct answer is (D).

8.   **D**   If (D) is a correct statement, then simply cutting out beef and lamb from your diet can dramatically decrease your carbon footprint—even if you do not decide to eat vegan. Choices (A), (B), and (C) are all true statements with respect to reducing the carbon footprint of food.

9.   **B**   A community that has been destroyed and then rebuilds is a result of secondary succession. Choice (A) would be the answer only if there was no life present before the new growth. Choice (C) is true only of a mature forest. Choice (D) is incorrect, as the species taking over the field is native to the rainforests on the periphery of the field. Nothing in the question suggests invasive species. Therefore, (B) is correct.

10.  **A**   If the intestines are occupied with beneficial bacteria, there will not be any room available for pathogenic bacteria to grow by density-dependent limitations on the population. Choice (B) is not accurate because pathogenic bacteria do not require a niche prepared for them by probiotic bacteria. Choice (C) is true, but it does not answer the question. Choice (D) is also incorrect, leaving (A) as your answer.

# Part VI
# Practice Test 2

# Practice Test 2

# AP® Biology Exam

## DO NOT OPEN THIS BOOKLET UNTIL YOU ARE TOLD TO DO SO.

### At a Glance

**Total Time**
1 hour and 30 minutes
**Number of Questions**
69
**Percent of Total Score**
50%
**Writing Instrument**
Pencil required

Instructions

Section I of this examination contains 69 multiple-choice questions. These are broken down into Part A (63 multiple-choice questions) and Part B (6 grid-in questions).

Indicate all of your answers to the multiple-choice questions on the answer sheet. No credit will be given for anything written in this exam booklet, but you may use the booklet for notes or scratch work. After you have decided which of the suggested answers is best, completely fill in the corresponding oval on the answer sheet. Give only one answer to each question. If you change an answer, be sure that the previous mark is erased completely. Here is a sample question and answer.

Sample Question          Sample Answer

Chicago is a

(A)    state
(B)    city
(C)    country
(D)    continent

Use your time effectively, working as quickly as you can without losing accuracy. Do not spend too much time on any one question. Go on to other questions and come back to the ones you have not answered if you have time. It is not expected that everyone will know the answers to all the multiple-choice questions.

**About Guessing**

Many candidates wonder whether or not to guess the answers to questions about which they are not certain. Multiple choice scores are based on the number of questions answered correctly. Points are not deducted for incorrect answers, and no points are awarded for unanswered questions. Because points are not deducted for incorrect answers, you are encouraged to answer all multiple-choice questions. On any questions you do not know the answer to, you should eliminate as many choices as you can, and then select the best answer among the remaining choices.

**This page intentionally left blank.**

# BIOLOGY
## SECTION I
### 69 Questions
### Time—90 minutes

Directions: Each of the questions or incomplete statements below is followed by four suggested answers or completions. Select the one that is best in each case and then fill in the corresponding oval on the answer sheet.

1. In general, animal cells differ from plant cells in that animal cells have

    (A) a cell wall made of cellulose
    (B) lysosomes
    (C) large vacuoles that store water
    (D) centrioles within centrosomes

2. A cell from the leaf of the aquatic plant Elodea was soaked in a 15 percent sugar solution, and its contents soon separated from the cell wall and formed a mass in the center of the cell. All of the following statements are true about this event EXCEPT

    (A) the vacuole lost water and became smaller
    (B) the space between the cell wall and the cell membrane expanded
    (C) the large vacuole contained a solution with much lower osmotic pressure than that of the sugar solution
    (D) the concentration of solutes in the extracellular environment is hypertonic with respect to the cell's interior

3. A chemical agent is found to denature all enzymes in the synaptic cleft. What effect will this agent have on acetylcholine?

    (A) Acetylcholine will not be released from the presynaptic membrane.
    (B) Acetylcholine will not bind to receptor proteins on the postsynaptic membrane.
    (C) Acetylcholine will not diffuse across the cleft to the postsynaptic membrane.
    (D) Acetylcholine will not be degraded in the synaptic cleft.

4. The base composition of DNA varies from one species to another. Which of the following ratios would you expect to remain constant in the DNA?

    (A) Cytosine : Adenine
    (B) Pyrimidine : Purine
    (C) Adenine : Guanine
    (D) Guanine : Deoxyribose

5. In reptile eggs, the extraembryonic membrane that functions in excretion and respiration is the

    (A) amnion
    (B) chorion
    (C) allantois
    (D) yolk sac

6. Consider the following enzyme pathway:

$$A \xrightarrow{1} B \xrightarrow{2} C \xrightarrow{3} D \xrightarrow{4} E \xrightarrow{5} F$$
$$\xdownarrow{6} X \xrightarrow{7} Y$$

    An increase in substance F leads to the inhibition of enzyme 3. All of the following are results of the process EXCEPT

    (A) an increase in substance X
    (B) increased activity of enzyme 6
    (C) decreased activity of enzyme 4
    (D) increased activity of enzyme 5

7. The liver is a vital organ that performs all of the following functions EXCEPT

    (A) storing amino acids that were absorbed in the capillaries of the small intestine
    (B) detoxifying harmful substances such as alcohol or certain drugs
    (C) synthesizing bile salts that emulsify lipids
    (D) breaking down peptides into amino acids

**GO ON TO THE NEXT PAGE.**

$pO_2$(mm Hg)

8. The graph above shows the oxygen dissociation curves of maternal hemoglobin and fetal hemoglobin. Based on the graph, it can be concluded that

    (A)  fetal hemoglobin surrenders $O_2$ more readily than maternal hemoglobin
    (B)  the dissociation curve of fetal hemoglobin is to the right of maternal hemoglobin
    (C)  fetal hemoglobin has a higher affinity for $O_2$ than does maternal hemoglobin
    (D)  fetal and maternal hemoglobin differ in structure

9. In minks, the gene for brown fur (B) is dominant over the gene for silver fur (b). Which set of genotypes represents a cross that could produce offspring with silver fur from parents that both have brown fur?

    (A)  BB × BB
    (B)  BB × Bb
    (C)  Bb × Bb
    (D)  Bb × bb

10. Hemoglobin is a molecule that binds to both $O_2$ and $CO_2$. There is an allosteric relationship between the concentrations of $O_2$ and $CO_2$. Hemoglobin's affinity for $O_2$

    (A)  decreases as blood pH decreases
    (B)  increases as $H^+$ concentration increases
    (C)  increases in exercising muscle tissue
    (D)  decreases as $CO_2$ concentration decreases

11. All viruses contain at least these two principal components:

    (A)  DNA and proteins
    (B)  nucleic acid and a capsid
    (C)  DNA and cell membrane
    (D)  RNA and cell wall

12. All of the following are differences between prokaryotes and eukaryotes EXCEPT

    (A)  eukaryotes have linear chromosomes, while prokaryotes have circular chromosomes
    (B)  eukaryotes possess double stranded DNA, while prokaryotes possess single stranded DNA
    (C)  eukaryotes process their mRNA, while in prokaryotes, transcription and translation occur simultaneously
    (D)  eukaryotes contain membrane-bound organelles, while prokaryotes do not

13. In humans, fertilization normally occurs in the

    (A)  ovary
    (B)  fallopian tube
    (C)  uterus
    (D)  placenta

14. The development of an egg without fertilization is known as

    (A)  meiosis
    (B)  parthenogenesis
    (C)  embryogenesis
    (D)  vegetative propagation

**GO ON TO THE NEXT PAGE.**

15. All of the following are examples of hydrolysis EXCEPT

    (A) conversion of fats to fatty acids and glycerol
    (B) conversion of proteins to amino acids
    (C) conversion of starch to simple sugars
    (D) conversion of pyruvic acid to acetyl CoA

16. In cells, which of the following can catalyze reactions involving hydrogen peroxide, provide cellular energy, and make proteins, in that order?

    (A) Peroxisomes, mitochondria, and ribosomes
    (B) Peroxisomes, mitochondria, and lysosomes
    (C) Peroxisomes, mitochondria, and Golgi apparatus
    (D) Lysosomes, chloroplasts, and ribosomes

17. All of the following play an important role in regulating respiration in humans EXCEPT

    (A) an increase in the amount of $CO_2$ in the blood
    (B) a decrease in the amount of $O_2$ in the blood
    (C) a decrease in the plasma pH level
    (D) strenuous exercise

18. The primary site of glucose reabsorption is the

    (A) glomerulus
    (B) proximal convoluted tubule
    (C) loop of Henle
    (D) collecting duct

Questions 19-20 refer to the following graph.

The graph below shows the growth curve of a bacterial culture.

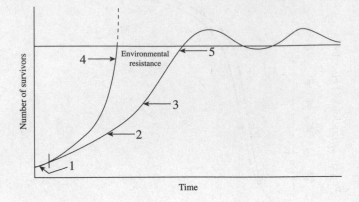

19. Which of the following represents the carrying capacity of the environment?

    (A) 2
    (B) 3
    (C) 4
    (D) 5

20. Which of the following shows the exponential growth curve of the population?

    (A) 1
    (B) 2
    (C) 3
    (D) 4

**GO ON TO THE NEXT PAGE.**

21. All of the following statements are true EXCEPT

    (A) thyroxine increases the rate of metabolism
    (B) insulin decreases storage of glycogen
    (C) vasopressin stimulates water reabsorption in the kidney
    (D) epinephrine increases blood sugar levels and heart rate

22. Metafemale syndrome, a disorder in which a female has an extra X chromosome, is the result of nondisjunction. The failure in oogenesis that could produce this would occur in

    (A) metaphase I
    (B) metaphase II
    (C) telophase I
    (D) anaphase II

23. In plants, the tendency of climbing vines to twine their tendrils around a trellis is called

    (A) thigmotropism
    (B) hydrotropism
    (C) phototropism
    (D) geotropism

24. Females with Turner's syndrome have a high incidence of hemophilia, a recessive, X-linked trait. Based on this information, it can be inferred that females with this condition

    (A) have an extra X chromosome
    (B) have an extra Y chromosome
    (C) lack an X chromosome
    (D) have red blood cells that clump

25. When a retrovirus inserted its DNA into the middle of a bacterial gene, it altered the normal reading frame by one base pair. This type of mutation is called

    (A) duplication
    (B) translocation
    (C) inversion
    (D) frameshift mutation

26. High levels of estrogen from maturing follicles inhibit the release of gonadotropin releasing hormone (GnRH). Which of the following endocrine glands produces GnRH?

    (A) Anterior pituitary
    (B) Posterior pituitary
    (C) Hypothalamus
    (D) Pineal gland

27. The principle inorganic compound found in living things is

    (A) carbon
    (B) oxygen
    (C) water
    (D) glucose

28. Kangaroo rats are better able to concentrate urine than humans are. It would be expected that, compared to the nephrons of human kidneys, the nephrons of kangaroo-rat kidneys would have

    (A) thicker walls, which are impermeable to water
    (B) shorter loops of Henle
    (C) longer loops of Henle
    (D) shorter collecting ducts

**GO ON TO THE NEXT PAGE.**

29. All of the following are modes of asexual reproduction EXCEPT

    (A) sporulation
    (B) fission
    (C) budding
    (D) meiosis

30. The moist skin of earthworms, spiracles of grasshoppers, and the mucus membranes lining alveoli are all associated with the process of

    (A) excretion
    (B) respiration
    (C) circulation
    (D) digestion

31. Invertebrate immune systems possess which of the following?

    (A) Cytotoxic T-lymphocytes
    (B) Phagocytes
    (C) B-cells
    (D) Helper T-cells

32. All of the following are examples of connective tissue EXCEPT

    (A) ligaments
    (B) muscle
    (C) blood
    (D) cartilage

33. Photosynthesis requires

    (A) glucose, light, $CO_2$
    (B) light, $CO_2$, water
    (C) water, soil, $O_2$
    (D) $O_2$, water, light

34. If a forest of fir, birch, and white spruce trees was devastated by fire, which of the following would most likely happen?

    (A) Only animal life would continue to inhabit the region.
    (B) Secondary succession would begin to occur.
    (C) Only tough grasses would appear.
    (D) The number of species would stabilize as the ecosystem matures.

35. Which of the following processes occur in the cytoplasm of an eukaryotic cell?

    I. DNA replication

    II. Transcription

    III. Translation

    (A) I only
    (B) III only
    (C) II and III only
    (D) I, II, and III

36. Crossing-over during meiosis permits scientists to determine

    (A) the chance for variation in zygotes
    (B) the rate of mutations
    (C) the distance between genes on a chromosome
    (D) which traits are dominant or recessive

37. An animal cell that is permeable to water but not salts has an internal NaCl concentration of 10%. If placed in freshwater the cell will

    (A) plasmolyze
    (B) swell and eventually lyse
    (C) endocytose water into a large central vacuole
    (D) shrivel

**GO ON TO THE NEXT PAGE.**

38. Three distinct bird species, flicker, woodpecker, and elf owl, all inhabit a large cactus, *Cereus giganteus*, in the desert of Arizona. Since competition among these birds rarely occurs, the most likely explanation for this phenomenon is that these birds

    (A) have a short supply of resources
    (B) have different ecological niches
    (C) do not live together long
    (D) are unable to breed

39. Lampreys attach to the skin of lake trout and absorb nutrients from its body. This relationship is an example of

    (A) commensalism
    (B) parasitism
    (C) mutualism
    (D) gravitropism

40. The nucleotide sequence of a DNA molecule is 5'-C-A-T-3'. A mRNA molecule with a complementary codon is transcribed from the DNA in the process of protein synthesis a tRNA pairs with a mRNA codon. What is the nucleotide sequence of the tRNA anticodon?

    (A) 5'-G-T-A-3'
    (B) 5'-G-U-A-3'
    (C) 5'-C-A-U-3'
    (D) 5-'U-A-C-3'

41. Viruses are considered an exception to the cell theory because they

    (A) are not independent organisms
    (B) have only a few genes
    (C) move about via their tails
    (D) have evolved from ancestral protists

42. All of the following organs in the digestive system secrete digestive enzymes EXCEPT the

    (A) mouth
    (B) stomach
    (C) gall bladder
    (D) small intestine

43. Memory loss would most likely be due to a malfunction of which part of the brain?

    (A) Medulla
    (B) Cerebellum
    (C) Cerebrum
    (D) Pons

**GO ON TO THE NEXT PAGE.**

44. The sequence of amino acids in hemoglobin molecules of humans is more similar to the hemoglobin of chimpanzees than it is to the hemoglobin of dogs. This similarity suggests that

    (A) humans and dogs are more closely related than humans and chimpanzees
    (B) humans and chimpanzees are more closely related than humans and dogs
    (C) humans are related to chimpanzees but not to dogs
    (D) humans and chimpanzees are closely analogous

45. According to the heterotroph hypothesis, which event had to occur before oxygen filled the atmosphere?

    (A) Heterotrophs had to remove carbon dioxide from the air.
    (B) Autotrophs, which make their own food, had to evolve.
    (C) Heterotrophs had to evolve.
    (D) Autotrophs had to convert atmospheric nitrogen to nitrate.

46. Two individuals, one with type B blood and one with type AB blood, have a child. The probability that the child has type O blood is

    (A) 0%
    (B) 25%
    (C) 50%
    (D) 100%

Questions 47–48 refer to the following bar graph, which shows the relative biomass of four different populations of a particular food pyramid.

Relative Biomass

47. The largest amount of energy is available to

    (A) population A
    (B) population B
    (C) population C
    (D) population D

48. Which of the following would be the most likely result if there was an increase in the number of organisms in population C?

    (A) The biomass of population D will remain the same.
    (B) The biomass of population B will decrease.
    (C) The biomass of population C will steadily increase.
    (D) The food source available to population C would increase.

**GO ON TO THE NEXT PAGE.**

Questions 49–52 refer to the following illustration and information.

The cell cycle is a series of events in the life of a dividing eukaryotic cell. It consists of four stages: $G_1$, S, $G_2$, and M. The duration of the cell cycle varies from one species to another and from one cell type to another. The $G_1$ phase varies the most. For example, embryonic cells can pass through the $G_1$ phase so quickly that it hardly exists, whereas neurons are arrested in the cell cycle and do not divide.

49. During which phase do chromosomes replicate?

   (A)  $G_1$
   (B)  S
   (C)  $G_2$
   (D)  M

50. In mammalian cells, the first sign of prophase is the

   (A)  appearance of chromosomes
   (B)  separation of chromatids
   (C)  disappearance of the nuclear membrane
   (D)  replication of chromosomes

51. Mitosis occurs in all of the following types of cells EXCEPT

   (A)  epidermal cells
   (B)  hair cells
   (C)  red blood cells
   (D)  pancreatic cells

52. Since neurons are destined never to divide again, what conclusion can be made?

   (A)  These cells will go through cell division.
   (B)  These cells will be permanently arrested in the $G_1$ phase.
   (C)  These cells will be permanently arrested in the $G_2$ phase.
   (D)  These cells will quickly enter the S phase.

**GO ON TO THE NEXT PAGE.**

Questions 53–56 refer to the following graphs, which show the permeability of ions during an action potential in a ventricular contractile cardiac fiber. The action potential of cardiac muscle fibers resembles that of skeletal muscles.

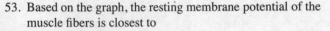

53. Based on the graph, the resting membrane potential of the muscle fibers is closest to

(A)  −90 mV
(B)  −70 mV
(C)  0 mV
(D)  +70 mV

54. Which of the following statements is true concerning the initial phase of depolarization?

(A)  Voltage-gated $K^+$ channels open in the plasma membrane.
(B)  The concentration of $Ca^{2+}$ ions within the plasma membrane becomes more negative.
(C)  The membrane potential stays close to −40 mV.
(D)  The permeability of the sarcolemma to $Na^+$ ions increases.

55. In cardiac fibers, the duration of an action potential is approximately

(A)  0.10 secs
(B)  0.20 secs
(C)  0.25 secs
(D)  0.30 secs

56. One major difference between the action potential of cardiac muscle fibers and the action potential of skeletal muscle fibers is that in cardiac muscle fibers

(A)  the membrane is permeable to $Na^+$, not $K^+$
(B)  voltage-gated $K^+$ channels open during depolarization, not repolarization
(C)  depolarization is prolonged compared to that in skeletal muscle fibers
(D)  the refractory period is shorter than that of skeletal muscle fibers

**GO ON TO THE NEXT PAGE.**

Questions 57–60 refer to the data below concerning the general animal body plan of five organisms.

| Characteristic | Sea anemone | Hagfish | Eel | Salamander |
|---|---|---|---|---|
| Vertebral column | | + | + | + |
| Jaws | + | | + | + |
| Walking legs | | | | + |

Note: + indicates a feature presence in an organism.

57. The two most closely related organisms are

    (A) sea anemone and hagfish
    (B) eel and salamander
    (C) hagfish and eel
    (D) sea anemone and salamander

58. The correct order of evolution for the traits above is

    (A) jaws – vertebral column – walking legs
    (B) walking legs – jaws – vertebral column
    (C) jaws – walking legs – vertebral column
    (D) vertebral column – jaws – walking legs

59. Pre- and post- zygotic barriers exist that prevent two different species from producing viable offspring. All of the following are pre-zygotic barriers EXCEPT

    (A) anatomical differences preventing copulation
    (B) different temporality of mating
    (C) sterility of offspring
    (D) incompatible mating songs

60. Birds and insects have both adapted wings to travel by flight. The wings of birds and insects are an example of

    (A) divergent evolution
    (B) convergent evolution
    (C) speciation
    (D) mutation

**GO ON TO THE NEXT PAGE.**

Questions 61–63 refer to the following synthetic pathway of a pyrimidine, cytidine 5' triphosphate, CTP. This pathway begins with the condensation of two small molecules by the enzyme, aspartate transcarbamylase (ATCase).

**Regulation of CTP biosynthesis**

61. Which of the following is true when the level of CTP is low in a cell?

   (A) CTP is converted to ATCase.
   (B) The metabolic traffic down the pathway increases.
   (C) ATCase is inhibited, which slows down CTP synthesis.
   (D) The final product of the pathway is reduced.

62. This enzymatic phenomenon is an example of

   (A) transcription
   (B) feedback inhibition
   (C) dehydration synthesis
   (D) photosynthesis

63. The biosynthesis of cytidine 5'-triphosphate requires

   (A) a ribose sugar, a phosphate group, and a nitrogen base
   (B) a deoxyribose sugar, a phosphate group, and a nitrogen base
   (C) a ribose sugar, phosphate groups, and a nitrogen base
   (D) a deoxyribose sugar, phosphate groups, and a nitrogen base

**GO ON TO THE NEXT PAGE.**

Directions: Part B consists of questions requiring numeric answers. Calculate the correct answer for each question.

64. In a diploid organism with the genotype AaBbCCDDEE, how many genetically distinct kinds of gametes would be produced?

65. Under favorable conditions, bacteria divide every 20 minutes. If a single bacterium replicated according to this condition, how many bacterial cells would one expect to find at the end of three hours?

**GO ON TO THE NEXT PAGE.**

66. In snapdragon plants that display intermediate dominance, the allele CR produces red flowers and CW produces white flowers. If a homozygous red-flowered snapdragon is crossed with a homozygous white-flowered snapdragon, what will the percentage of pink offspring be?

67. Translation is an energy-intensive process. Approximately how many ATPs are required to synthesize a protein containing 115 amino acids?

**GO ON TO THE NEXT PAGE.**

Question 68 refers to the following experiment.

A group of 100 *Daphnia*, small crustaceans known as water fleas, were placed in one of three culture jars of different sizes to determine their reproductive rate. The graph below shows the average number of offspring produced per female each day in each jar of pond water.

Key: Ⓐ Water fleas in a 1-liter jar of pond water
Ⓑ Water fleas in a 0.5-liter jar of pond water
Ⓒ Water fleas in a 0.25-liter jar of pond water

68. What is the total number of offspring produced in the 0.5-liter jar on the twentieth day, assuming all survive?

**GO ON TO THE NEXT PAGE.**

69. On average, there is a 90 percent reduction of productivity for each trophic level. Based on this information, 10,000 pounds of grass should be able to support how many pounds of crickets?

**END OF SECTION I**

# BIOLOGY
## SECTION II

**8 Questions**
**Planning Time—10 minutes**
**Writing Time—80 minutes**

Directions: Questions 1 and 2 are long free-response questions that should require about 22 minutes each to answer and are worth 10 points each. Questions 3 through 8 are short free-response questions that should require about 6 minutes each to answer. Questions 3 through 5 are worth 4 points each, and questions 6 through 8 are worth 3 points each.

Read each question carefully and completely. Write your response in the space provided following each question. Only material written in the space provided will be scored. Answers must be written out in paragraph form. Outlines, bulleted lists, or diagrams alone are not acceptable.

1. Chlorophyll is one of a class of pigments that absorbs light energy in photosynthesis.

   (a) **Relate** the function of chlorophyll to its location.

   (b) **Design** an experiment to investigate the influence of sunlight on the activity of chlorophyll.

   (c) **Describe** what information concerning the structure of chlorophyll could be inferred from your experiment.

2. Over the course of early evolution, organisms had to develop various methods to regulate internal fluids and excrete wastes. **Discuss** the problems faced by **three** organisms and how these problems were solved. In your discussion include structural adaptations and their functional significance.

3. **Describe** the chemical nature of genes. Name two types of gene mutations that could occur during replication.

4. **Select one** of the following three pairs of hormones and discuss the concept of negative feedback.

   (a) Thyroid-stimulating hormone (TSH) and thyroxine

   (b) Parathyroid hormone and calcitonin

   (c) ACTH and cortisol

5. **Describe** why fermentation is a less efficient way to produce energy than aerobic respiration.

6. **Define** analogous structures and give an example.

7. **Describe** symbiosis and give an example involving humans.

8. **Define** the role of ADH (anti-diuretic hormone) and aldosterone in the regulation of blood pressure.

## STOP

## END OF EXAM

# Practice Test 2:
# Answers and Explanations

# PRACTICE TEST 2 ANSWER KEY

| | | | |
|---|---|---|---|
| 1. D | | 36. C | |
| 2. D | | 37. B | |
| 3. D | | 38. B | |
| 4. B | | 39. B | |
| 5. C | | 40. C | |
| 6. D | | 41. A | |
| 7. D | | 42. C | |
| 8. C | | 43. C | |
| 9. C | | 44. B | |
| 10. A | | 45. B | |
| 11. B | | 46. A | |
| 12. A | | 47. B | |
| 13. B | | 48. B | |
| 14. B | | 49. B | |
| 15. D | | 50. A | |
| 16. A | | 51. C | |
| 17. B | | 52. B | |
| 18. B | | 53. A | |
| 19. D | | 54. D | |
| 20. D | | 55. D | |
| 21. B | | 56. C | |
| 22. D | | 57. B | |
| 23. A | | 58. D | |
| 24. C | | 59. C | |
| 25. D | | 60. B | |
| 26. C | | 61. B | |
| 27. C | | 62. B | |
| 28. C | | 63. C | |
| 29. D | | 64. 4 | |
| 30. B | | 65. 512 | |
| 31. B | | 66. 100 | |
| 32. B | | 67. 460 | |
| 33. B | | 68. 200 | |
| 34. B | | 69. 1,000 | |
| 35. B | | | |

# PRACTICE TEST 2 EXPLANATIONS

## Section I: Multiple Choice

1. **D**    Animal cells have centrioles. Both animal and plant cells have an endoplasmic reticulum, membrane-bound organelles, and lysosomes. Only plants have a cell wall made of cellulose and large vacuoles.

2. **D**    If the contents of the cell separated from the cell wall, then water was moving out of the cell. This would cause the space between the cell wall and the cell membrane to expand, so eliminate (B). This also means that the concentration of solutes in the extracellular environment would be hypotonic with respect to the cell's interior, which is why (D) is the correct answer. You can eliminate (C) because the vacuole is within the cell; it contains a solution with a much lower osmotic pressure than that of the sugar solution. Lastly, eliminate (A): Because the fluid in the cell was hypertonic to the sugar solution, fluid was moving out of the vacuole and caused it to become smaller.

3. **D**    This question tests your ability to associate what happens when enzymes are denatured and what would happen in the synaptic cleft. Acetylcholinesterase is an enzyme that degrades acetylcholine in the synaptic cleft. If acetylcholinesterase is denatured, acetylcholine will still be released from the presynaptic membrane and continue to diffuse across the synaptic cleft and bind to the postsynaptic membrane because acetylcholine is not degraded. Therefore, you can eliminate (A), (B), and (C).

4. **B**    The ratio of purines to pyrimidines should be constant because purines always bind with pyrimidines, no matter which ones they may be.

5. **C**    The function of the allantois is gas exchange, so the correct answer is (C). Choice (A), the amnion, is the fluid-filled sac that protects the embryo. Choice (B), the chorion, is the outermost membrane, and (D), the yolk sac, is the food for the embryo.

6. **D**    If substance F leads to the inhibition of enzyme 3, then substances D and E and enzymes 3, 4, and 5 will be affected. The activity of enzyme 5 will be decreased, not increased.

7. **D**    The liver does not break down peptides into amino acids. It performs all of the following functions: stores amino acids absorbed in the capillaries, as in (A); detoxifies harmful substances, as in (B); and makes bile, as in (C).

8. **C**    Based on the graph, fetal hemoglobin has a higher affinity for oxygen than maternal hemoglobin. Fetal hemoglobin does not give up oxygen more readily than maternal hemoglobin, so eliminate (A). You can also get rid of (B), as the dissociation curve of fetal hemoglobin is to the left of the maternal hemoglobin. Finally, eliminate (D) because fetal hemoglobin and maternal hemoglobin are different structurally, but you can't tell this from the graph.

9. **C**    The set of genotypes that represents a cross that could produce offspring with silver fur from parents that both have brown fur is Bb and Bb. Complete a Punnett square for this question. In order for the offspring to have silver fur, both parents must have the silver allele.

10. **A** Choices (B) and (D) are incorrect: Hemoglobin's affinity for $O_2$ decreases as the concentration of $H^+$ increases (or the pH decreases) and as the concentration of $CO_2$ increases (or the concentration of $HCO_3^-$ increases). Also eliminate (C), as hemoglobin's affinity for oxygen in tissue muscles does not increase during exercise.

11. **B** Viruses are made up of nucleic acid surrounded by a protein coat called a capsid. They do not contain a cell wall, proteins, or cell membrane.

12. **A** Both prokaryotes and eukaryotes possess double-stranded DNA. Choices (B), (C), and (D) are correct differences between prokaryotes and eukaryotes.

13. **B** In humans, fertilization normally occurs in the fallopian tube. The ovary is the female gonad that contains eggs, so eliminate (A). The uterus is the organ that houses the developing embryo, so eliminate (C). Finally, (D) can also be eliminated because the placenta is the structure that nourishes the embryo.

14. **B** The development of an egg without fertilization is known as parthenogenesis. Parthenogenesis is a form of asexual reproduction found in insects and lizards. Choice (A), meiosis, is a form of sexual reproduction that produces gametes. Choice (C), embryogenesis, refers to the early stages of embryo development. Choice (D), vegetative propagation, is a form of asexual reproduction by which plants produce identical offspring from stem, leaves, or roots.

15. **D** All of the choices are examples of hydrolysis except the conversion of pyruvic acid to glucose. Hydrolysis is the breaking of a covalent bond by adding water. In all of the correct examples, complex compounds are broken down to simpler compounds. The conversion of pyruvic acid to acetyl CoA is an example of decarboxylation—a carboxyl group is removed as carbon dioxide and the 2-carbon fragment is oxidized.

16. **A** Peroxisomes catalyze reactions that produce hydrogen peroxide, ribosomes are involved in protein synthesis, and mitochondria contain enzymes involved in cellular respiration. Eliminate (B) and (D) because lysosomes are the sites of degradation; they contain hydrolytic enzymes but do not produce hydrogen peroxide. Choice (C) is incorrect, as the Golgi apparatus sorts and packages substances that are destined to be secreted out of the cell. Interestingly, $O_2$ concentration generally doesn't play an important role in regulating respiration.

17. **B** The respiratory rate in humans will be affected by an increase in the amount of $CO_2$ and a drop in pH levels, which is the same as an increase in hydrogen ion levels. Strenuous exercising will also modify the respiratory rate. $O_2$ concentration generally does not play an important role in regulating respiration.

18. **B** The primary site of glucose reabsorption is the proximal convoluted tubule. The glomerulus (A) is a tuft of capillaries that filters fluid into the Bowman's capsule. Choice (C), the loop of Henle, is the site of salt reabsorption. Choice (D), the collecting duct, is the site in which urine is concentrated.

19. **D** The carrying capacity is the maximum number of organisms of a given species that can be maintained in a given environment. Once a population reaches its carrying capacity, the number of organisms will fluctuate around it.

20. **D** During the exponential growth phase of a population, the size doubles during each time interval. This part of the graph looks like a parabola.

21. **B** Insulin decreases the level of blood glucose by increasing the storage of glycogen in muscles. The other hormones are correctly paired with their respective function.

22. **D** The failure in oogenesis that could produce this syndrome would occur in anaphase I or anaphase II. Anaphase refers to the stage of meiosis in which chromatids separate from each other. If the chromosomes or chromatids fail to separate during anaphase, one egg cell will contain two X chromosomes instead of one.

23. **A** The tendency of climbing vines to twine their tendrils around a trellis is called thigmotropism. Thigmotropism refers to growth stimulated by contact with an object. Choice (B), hydrotropism, refers to growth of a plant toward water. Choice (C), phototropism, refers to growth toward light. Choice (D), geotropism, refers to growth toward or against gravity.

24. **C** Females with Turner's syndrome lack an X chromosome. If females with this syndrome have a high rate of hemophilia, they must not have the second X to mask the expression of the disease.

25. **D** This type of mutation is called frameshift mutation. The insertion of DNA leads to a change in the normal reading frame by one base pair. The other answer choices refer to chromosomal aberrations. Choice (A), duplication, is when an extra copy of a chromosome segment is introduced. Choice (B), translocation, is when a segment of a chromosome moves to another chromosome. Choice (C), inversion, is when a segment of a chromosome is inserted in the reverse orientation.

26. **C** The hypothalamus releases GnRH, which stimulates the release of FSH. The key word in this question is "releasing." The organ that produces releasing or inhibiting factors is the hypothalamus. Choice (A), the anterior pituitary, is the master gland that secretes several hormones. Choice (B), the posterior pituitary, secretes oxytocin and vasopressin. Choice (D), the pineal gland, is a gland at the base of the brain that secretes melatonin and helps regulate circadian rhythms.

27. **C** The principle inorganic compound found in living things is water. Water is a necessary component for life.

28. **C** The loop of Henle is responsible for the concentration of urine. The longer the loop of Henle, the more water would be reabsorbed, which would make the filtrate more concentrated. Eliminate (D); if the loop of Henle were longer, the collecting ducts may or may not also be longer. Choice (A) is also incorrect; if the walls of the nephrons were thicker and impermeable, water would not be conserved.

29. **D** All of the following are modes of asexual reproduction except meiosis, which is the production of gametes for sexual reproduction. Choice (A), sporulation, is a form of asexual reproduction in which spores are produced. Choice (B), fission, is the equal division of a bacterial cell. Choice (C), budding, is a form of asexual reproduction in yeasts in which small cells grow from a parent cell.

30. **B** Respiration is the transport of oxygen into an organism and carbon dioxide out of an organism.

31. **B**     Invertebrates lack specific immune responses including B-cells and T-cells. Phagocytes is the only answer choice that describes non-specific immune responses.

32. **B**     All of the choices are examples of connective tissue except muscle. Connective tissue connects and supports other tissues. Ligaments, blood, cartilage, and bone are all connective tissues.

33. **B**     Photosynthesis requires light, carbon dioxide, and water and produces oxygen and glucose.

34. **B**     When a forest of trees is devastated by fire, secondary succession is most likely to occur. Secondary succession refers to ecological succession in a disturbed community. Choice (A) is incorrect because plants and animals will continue to inhabit the region once the community is reestablished. You can also eliminate (C), as tough grasses are not pioneers. Finally, (D) is incorrect because it cannot be determined if the number of species will be stabilized.

35. **B**     Translation, the synthesis of proteins from mRNA, occurs in the cytoplasm. DNA replication (I) occurs in the nucleus. Transcription (II), the synthesis of RNA from DNA, occurs in the nucleus.

36. **C**     Crossing-over permits scientists to determine chromosome mapping. Chromosome mapping is a detailed map of all the genes on a chromosome. The frequency of crossing-over between any two alleles is proportional to the distance between them. The farther apart the two linked alleles are on a chromosome, the more often the chromosome will break between them. Crossing-over does not tell us about the chance of variation in zygotes, the rate of mutations, or whether the traits are dominant, recessive, or masked.

37. **B**     In this scenario, the inside of the cell is hypertonic to the outside environment—this will cause water to move into the cell and can cause the cell to lyse.

38. **B**     The most likely explanation for this phenomenon is that these birds have different ecological niches. An ecological niche is the position or function of an organism or population in its environment. Eliminate (A) because we do not know if there is a short supply of resources. Choice (C) can also be eliminated because we do not know how long the bird species live together. The breeding patterns of the bird species does not explain the lack of competition, which eliminates (D) as well.

39. **B**     This relationship is an example of parasitism. Parasitism is a form of symbiosis in which one organism benefits and the other is harmed. Choice (A), commensalism, is a form of symbiosis in which one organism benefits and the other is unaffected. Choice (C), mutualism, is a form of symbiosis in which both organisms benefit. Choice (D), gravitropism, is the growth of a plant toward or away from gravity.

40. **C**     If the nucleotide sequence of a DNA molecule is 5'-C-A-T-3', then the transcribed DNA strand (mRNA) would be 3'-G-U-A-5'. The nucleotide sequence of the tRNA codon would be 5'-C-A-U-3'.

41. **A**     Viruses are considered an exception to the cell theory because they are not independent organisms. They can only survive by invading a host. Choice (B) is incorrect, as viruses have either DNA or RNA. Choice (C) is also wrong because not all viruses have a tail. You can also eliminate (D), as viruses did not evolve from ancestral protists.

42.  **C**  All of the listed organs secrete digestive enzymes except the gall bladder. The gall bladder stores and secretes bile produced by the liver, which is an emulsifier, not an enzyme. The mouth, (A), secretes salivary amylase. The stomach, (B), secretes pepsin. The small intestine, (D), secretes protease.

43.  **C**  Memory loss would most likely be due to a malfunction of the cerebrum. The cerebrum controls all voluntary activities and receives and interprets sensory information. The medulla, (A), controls involuntary actions such as breathing. The cerebellum, (B), coordinates muscle activity and controls balance. The pons, (D), is a mass of nerve fibers running across the surface of the mammalian brain.

44.  **B**  The similarity suggests that humans and chimpanzees are more closely related than humans and dogs. Because these two organisms share similar amino acid sequences, humans must share more recent common ancestors with chimpanzees than with dogs.

45.  **B**  The event that had to occur before oxygen filled the atmosphere was that autotrophs, which make their own food and give off oxygen, had to evolve.

46.  **A**  There is no way for these two parents to produce a type O child. The only genotype that will result in type O blood is two "O" alleles, one from each parent, since the allele for O blood is recessive to the alleles for A or B blood. The AB parent only has alleles for A and B blood, so it is impossible for these two individuals to produce a child with type O blood.

47.  **B**  The largest amount of energy is available to producers. Population B is most likely composed of producers because they have the largest biomass.

48.  **B**  An increase in the number of organisms in population C would most likely lead to a decrease in the biomass of B because population B is the food source for population C. Make a pyramid based on the biomasses given. If population C increases, population B will decrease. Eliminate (A) and (C), as we cannot necessarily predict what will happen to the biomass of populations that are above population C. Choice (D) can also be eliminated because the food source available to population C would most likely decrease, not increase.

49.  **B**  Chromosomes replicate during interphase, the S phase. Choices (A) and (C) are incorrect because during $G_1$ and $G_2$, the cell makes protein and performs other metabolic duties.

50.  **A**  The first sign of prophase in mammalian cells is the appearance of chromosomes.

51.  **C**  Mitosis occurs in all of the following type of cells except mature red blood cells. Mature red blood cells are short-lived and do not divide.

52.  **B**  Because neurons are not capable of dividing, it is reasonable to conclude that these cells will not complete the $G_1$ phase. This is a reading comprehension question. The passage states that cells that do not divide are arrested at the $G_1$ phase. Choice (A) is incorrect because these cells will not be committed to go through cell division. You can also eliminate (C) and (D), as the cells will not enter the $G_2$ or S phase.

53.  **A**  According to the graph, the resting membrane potential of the muscle fiber is close to −90mV.

54. **D** Refer to the second graph about the membrane permeability of ions during a muscle contraction. During depolarization, the membrane is permeable to $Na^+$. Choice (A) can be eliminated because the voltage-gated $K^+$ channels do not open until after depolarization. Eliminate (B) because the concentration of $Ca^{2+}$ does not become more negative. Choice (C) is also incorrect, as the membrane potential changes from –90 mV to +20 mV.

55. **D** Refer to both graphs in the passage. In cardiac fibers, the duration of an action potential—a neuronal impulse—is approximately 0.3 seconds.

56. **C** In cardiac muscle fibers, depolarization is prolonged compared to that in skeletal muscle fibers. Eliminate (A) because the membrane is permeable to both $Na^+$ and $K^+$. Choice (B) can also be eliminated; in cardiac muscle fibers, voltage-gated $K^+$ channels open during repolarization. Finally, (D) is wrong because the refractory period is longer in cardiac muscle fibers compared to skeletal muscle fibers.

57. **B** The two most closely related organisms are the two with the most shared derived characteristics.

58. **D** Shared derived characteristics are newly evolved traits that are shared with every group on a phylogenic tree except for one. Vertebral columns are present in every group except for the sea anemone, so it must have evolved first. Walking legs are only found in the salamander, indicating that it most likely evolved most recently.

59. **C** Pre-zygotic barriers to reproduction are those that prevent fertilization, so you can eliminate (A), (B), and (D). Choice (C) is an example of a post-zygotic barrier to reproduction.

60. **B** Convergent evolution occurs when two organisms that are not closely related independently evolve similar traits, such as the wings of insects and birds. Divergent evolution occurs when two closely related individuals become more different over time and can lead to speciation.

61. **B** When the level of CTP is low in a cell, the metabolic traffic down the pathway increases. This pathway is controlled by feedback inhibition. The final product of the pathway inhibits the activity of the first enzyme. When the supply of CTP is low, the pathway will continue to produce CTP.

62. **B** This enzymatic phenomenon is an example of feedback inhibition. Feedback inhibition is the metabolic regulation in which high levels of an enzymatic pathway's final product inhibit the activity of its rate-limiting enzyme. Transcription, (A), is the production of RNA from DNA. Dehydration synthesis, (C), is the formation of a covalent bond by the removal of water. In photosynthesis, (D), radiant energy is converted to chemical energy.

63. **C** The biosynthesis of cytidine 5'-triphosphate requires a nitrogenous base, three phosphates, and the sugar ribose. Pyrimidines are a class of nitrogenous bases with a single ring structure. The sugar they contain is ribose, which is shown in the pathway diagram.

64. **4** There are four genetically distinct kinds of gametes that could be produced, ABCDE, AbCDE, aBCDE, and abCDE. Notice that the only alleles that vary are A and B ($2^2 = 4$).

65. **512**   If bacteria divide every 20 minutes, you would produce 512 bacterial cells. One method would be to use the equation $2^x$, where $x$ equals the number of 20 minute intervals in three hours; $2^9 = 512$.

66. **100**   Because snapdragon plants display intermediate dominance, the heterozygous phenotype is affected by the alleles of both homozygotes. If a homozygous red plant is crossed with a homozygous white plant, all offspring would be heterozygous and pink.

67. **460**   To calculate an estimate of how many ATPs are used in the synthesis of a protein, the number of amino acids in the protein should be multiplied by four. Since the question states there are 115 amino acids, the answer is 460.

68. **200**   The graph shows the average number of offspring per female per day. Because there are 100 females in the $\frac{1}{2}$ liter, the total number of offspring on the twentieth day would be 100 females times 2 offspring per day, which equals 200 offspring.

69. **1,000**   A 90 percent reduction of productivity in the grass is a reduction of 9,000 pounds. That means the grass should be able to support 1,000 pounds of crickets.

# Section II: Free Response

## Sample Long-Form Free Response 1

Chlorophyll is the principal light-harnessing pigment found in the thylakoid membrane of chloroplasts. It is a molecule that's made up of a large ring structure composed of smaller rings (pyrroles). In the center of the large porphyrin ring is a magnesium atom surrounded by four nitrogen atoms. The small rings consist of many alternating single and double bonds. These alternating bonds allow electrons in the porphyrin ring and magnesium atoms to move around freely. Alternating double and single bonds are commonly found in molecules that strongly absorb visible light. Photosynthesis occurs when sunlight activates chlorophyll by exciting their electrons. The chlorophyll molecule has a long hydrophobic hydrocarbon tail. This shape allows many chlorophyll molecules to be grouped together like a stack of saucers.

The following experiment could determine the influence of sunlight on chlorophyll. First, extract chlorophyll from the leaf by boiling it gently in a dissolved solution. Then pass light of a known wavelength through the solution and measure it, using a spectrophotometer. The wavelength of the entering light can be varied to see which wavelength is most absorbed by the solution. You could then plot the data on a graph to form an absorption spectrum. The absorption spectrum will show two peaks, which represent the type of light absorbed by the pigment. The valley would represent the light that's reflected.

Based on this experiment, it would be inferred that certain wavelengths, such as orange-red and violet-blue, are strongly absorbed by chlorophyll, whereas green and yellow hues are least absorbed, and are reflected. Chlorophyll gives plants their characteristic green color because it reflects yellow-green light.

## Sample Long-Form Free Response 2

Because the amount of intracellular water is critical, one major challenge that various organisms faced was how they would regulate body water and get rid of wastes. Waste products include carbon dioxide, salts, and nitrogenous wastes. This issue is especially important because nitrogenous wastes are highly toxic to the body. Over the course of evolution, organisms developed various structural adaptations to deal with this problem.

Insects developed excretory organs called Malpighian tubules. These long, slender tubules take up water and salt, concentrate the waste, and empty it into the intestine. The waste product, uric acid, is excreted as a dry pellet through the anus. Insects also have a hard, dry cuticle that has a waxy outer layer; this aids in preventing water loss.

Organisms that live in the sea, such as marine fishes, had to cope with the gradual increase in the saltiness of the water. Seawater is hypertonic relative to their body fluids, and marine fish had to protect their body cells from water loss because without a specific mechanism for this, they would become dehydrated even though they're surrounded by water. They excrete concentrated urine that is isotonic with their body fluids. They also eliminate excess salt by actively pumping it out through their gills.

Because humans are terrestrial organisms, they must conserve plenty of water. Humans also must get rid of nitrogenous wastes, and their major excretory organ is the kidney. Each kidney is made up of millions of function units called nephrons. As the filtrate travels along the nephron, glucose, amino acids, and salts are retained by the body, and the rest of the fluid is concentrated into urine. The skin is another important organ that gets rid of wastes. It contains sweat glands that help to maintain an optimal salt balance in the body.

## Sample Short-Form Free Response 3

A gene is a heredity unit located at a specific locus along a chromosome. Genes are made up of DNA, and DNA is made up of repeating subunits of nucleotides. A nucleotide consists of three parts, a five-carbon sugar, a phosphate group, and a nitrogenous base.

A gene mutation is a change in the sequence of base pairs in a DNA. It results from defects in the sequence of bases. Mutations that involve a single base change in the DNA sequence are called point mutations. These are base substitutions involving a single DNA nucleotide being replaced by another nucleotide. Another type of mutation is a frameshift mutation, caused by an insertion or deletion of bases in the DNA sequence. This causes a shift in the reading frame of DNA.

## Sample Short-Form Free Response 4

Hormones are chemical messengers that travel through the blood and act on target cells in the body. Hormone secretion is regulated by a negative feedback system, which is the basis of hormonal regulation. In part, negative feedback loops mean that an excess of a hormone in the bloodstream temporarily shuts down the production of that hormone.

The anterior pituitary gland secretes a hormone called thyroid-stimulating hormone (TSH). TSH stimulates the thyroid gland to secrete its hormone; thyroxine. Thyroxine regulates the basal metabolic rate in most body tissues. If the blood level of thyroxine rises above normal, it will suppress the secretion of TSH which, in turn, will lower the blood level of thyroxine. Both responses are negative feedback mechanisms.

Two hormones play a critical role in the regulation of calcium, parathyroid hormone and calcitonin. Parathyroid hormone, which is secreted from the parathyroid glands, increases blood calcium levels. It triggers the bones to release the calcium stored inside them. Calcitonin acts as an antagonist of parathyroid hormone. It decreases blood calcium levels. When blood calcium levels are above normal, the thyroid gland secretes calcitonin. This mechanism helps to maintain homeostasis.

ACTH is another hormone that's released by the anterior pituitary. It stimulates the adrenal cortex to secrete a number of steroid hormones, including cortisol. Cortisol increases the blood's concentration of glucose and helps the body adapt to stress. When cortisol levels reach a peak level in the bloodstream, the anterior pituitary is temporarily prevented from producing ACTH. This, in turn, shuts down the production of cortisol. Once the level of cortisol falls below normal, the adrenal cortex resumes production of cortisol.

## Sample Short-Form Free Response 5

Fermentation occurs when oxygen is not available to act as the final electron acceptor. When this does not occur, then NADH and $FADH_2$ are not able to be regenerated into $NAD^+$ and FAD, respectively. Instead, the only step in respiration that is occurring is glycolysis, thus generating its end product of pyruvate. Pyruvate can act as the final electron acceptor and allow $NAD^+$ to be regenerated, but the only energy that has been produced is the net of two ATP from glycolysis. Since not all the energy can be mined from the carbons when processed in fermentation, the process is less efficient.

## Sample Short-Form Free Response 6

Analogous structures are ones which have similar functions, but whose underlying structures are very different. They are often the product of convergent evolution, where differing organisms ended up in the same environment and thus subjected to the same forces of natural selection. An example of analogous structures would be the wing of bird and the wing of a bee. Both are used for flight, but their components are completely different.

## Sample Short-Form Free Response 7

Symbiosis describes a situation in which two organisms co-exist, in which at least one benefits by gaining food, shelter, transportation or some other requirement for existence. If the host provides a benefit, but remains unimpacted, then it is commensalistic, but if both organisms benefit then it is mutualistic. An example involving humans would be the bacterial flora of the skin or of the intestine. Many of these bacteria provide a protective or digestive benefit to humans while some utilize the body for food while also not doing any harm.

## Sample Short-Form Free Response 8

When blood pressure drops, aldosterone is released and allows more sodium to be reabsorbed into the blood stream at the distal convoluted tubule of the nephron of the kidney. When the concentration of the blood goes up, it is possible for it to take up more fluid so when aldosterone acts on the kidney, it is followed by anti-diuretic hormone. ADH creates water channels which then allow water to pass from the collecting duct into the circulatory system; this water is following the sodium that has already passed into the blood. By increasing the volume in the circulatory system, the body's blood pressure is raised.

# About the Author

Kim Magloire received a degree in Biology and Science in Human Affairs from Princeton University. She received her Master's degree in Epidemiology from Columbia University and is completing her doctorate in Epidemiology. In her many years of teaching, Kim has prepared hundreds of students for the AP Biology Exam as well as for the SAT, MCAT, GMAT, LSAT, GRE, and other science SAT Subject Tests.

Kim is the founder of SciTech, an educational company dedicated to helping students achieve academic excellence in science, math, and technology. Founded in 1996, SciTech offers a number of programs that help students make the "science and math connection." She has received numerous awards for her innovative educational programs in New York City, including the 1999 National "Community Leader of the Year" award from The Quaker Oats Company. She resides in New York City with her husband and daughter.

# AP BIOLOGY
# EQUATIONS AND FORMULAS

## Statistical Analysis and Probability

### Mean

$$\bar{x} = \frac{1}{n}\sum_{i=1}^{n} x_i$$

### Standard Deviation*

$$S = \sqrt{\frac{\sum (x_i - \bar{x})^2}{n-1}}$$

### Standard Error of the Mean*

$$SE_{\bar{x}} = \frac{S}{\sqrt{n}}$$

### Chi-Square

$$x^2 = \sum \frac{(o-e)^2}{e}$$

$\bar{x}$ = sample mean

$n$ = size of the sample

$s$ = sample standard deviation (i.e., the sample-based estimate of the standard deviation of the population)

$o$ = observed results

$e$ = expected results

Degrees of freedom are equal to the number of distinct possible outcomes minus one.

### Chi-Square Table

| $p$-value | Degrees of Freedom | | | | | | | |
|---|---|---|---|---|---|---|---|---|
| | 1 | 2 | 3 | 4 | 5 | 6 | 7 | 8 |
| 0.05 | 3.84 | 5.99 | 7.82 | 9.49 | 11.07 | 12.59 | 14.07 | 15.51 |
| 0.01 | 6.64 | 9.21 | 11.34 | 13.28 | 15.09 | 16.81 | 18.48 | 20.09 |

### Laws of Probability

If A and B are mutually exclusive, then:

$$P(A \text{ or } B) = P(A) + P(B)$$

If A and B are independent, then:

$$P(A \text{ and } B) = P(A) \times P(B)$$

### Hardy-Weinberg Equations

$p^2 + 2pq + q^2 = 1$    $p$ = frequency of the dominant allele in a population

$p + q = 1$    $q$ = frequency of the recessive allele in a population

### Metric Prefixes

| Factor | Prefix | Symbol |
|---|---|---|
| $10^9$ | giga | G |
| $10^6$ | mega | M |
| $10^3$ | kilo | k |
| $10^{-2}$ | centi | c |
| $10^{-3}$ | milli | m |
| $10^{-6}$ | micro | $\mu$ |
| $10^{-9}$ | nano | $n$ |
| $10^{-12}$ | pico | $p$ |

Mode = value that occurs most frequently in a data set

Median = middle value that separates the greater and lesser halves of a data set

Mean = sum of all data points divided by the number of data points

Range = value obtained by subtracting the smallest observation (sample minimum) from the greatest (sample maximum)

*For the purposes of the AP Exam, students will not be required to perform calculations using this equation; however, they must understand the underlying concepts and applications.*

## Rate and Growth

### Rate

$$\frac{dY}{dt}$$

### Population Growth

$$\frac{dN}{dt} = B - D$$

### Exponential Growth

$$\frac{dN}{dt} = {}^r\!\max{}^N$$

### Logistic Growth

$$\frac{dN}{dt} = {}^r\!\max{}^N\left(\frac{K - N}{K}\right)$$

### Temperature Coefficient Q$_{10}$‡

$$Q_{10} = \left(\frac{k_2}{k_1}\right)^{\frac{10}{T_2 - T_1}}$$

### Primary Productivity Calculation

$$\frac{mg\,O_2}{L} \times \frac{0.698\,mL}{mg} = \frac{mL\,O_2}{L}$$

$$\frac{mL\,O_2}{L} \times \frac{0.536\,mg\,C\ fixed}{mL\,O_2} = \frac{mg\,C\ fixed}{L}$$

(at standard temperature and pressure)

---

$dY$ = amount of change

$dt$ = change in time

$B$ = birth rate

$D$ = death rate

$N$ = population size

$K$ = carrying capacity

$r_{max}$ = maximum per capita growth rate of population

---

$T_2$ = higher temperature

$T_1$ = lower temperature

$k_2$ = reaction rate at $T_2$

$k_1$ = reaction rate at $T_1$

$Q_{10}$ = the factor by which the reaction rate increases when the temperature is raised by ten degrees

---

## Water Potential (ψ)

$$\psi = \psi_P + \psi_S$$

$\psi_P$ = pressure potential

$\psi_S$ = solute potential

The water potential will be equal to the solute potential of a solution in an open container because the pressure potential of the solution in an open container is zero.

### The Solute Potential of a Solution

$$\psi_S = -iCRT$$

$i$ = ionization constant (this is 1.0 for sucrose because sucrose does not ionize in water)

$C$ = molar concentration

$R$ = pressure constant ($R = 0.0831$ liter bars/mole K)

$T$ = temperature in Kelvin (°C + 273)

---

## Surface Area and Volume

### Volume of a Sphere

$$V = \frac{4}{3}\pi r^3$$

### Volume of a Rectangular Solid

$$V = lwh$$

### Volume of a Right Cylinder

$$V = \pi r^2 h$$

### Surface Area of a Sphere

$$A = 4\pi r^2$$

### Surface Area of a Cube

$$A = 6s^2$$

### Surface Area of a Rectangular Solid

$$A = \sum \text{surface area of each side}$$

---

$r$ = radius

$l$ = length

$h$ = height

$w$ = width

$s$ = length of one side of a cube

$A$ = surface area

$V$ = volume

$\sum$ = sum of all

---

## Dilution (used to create a dilute solution from a concentrated stock solution)

$$C_i V_i = C_f V_f$$

i = initial (starting)     $C$ = concentration of solute

f = final (desired)     $V$ = volume of solution

---

## Gibbs Free Energy

$$\Delta G = \Delta H - T\Delta S$$

$\Delta G$ = change in Gibbs free energy

$\Delta S$ = change in entropy

$\Delta H$ = change in enthalpy

$T$ = absolute temperature (in Kelvin)

---

$$pH^* = -\log_{10}[H^+]$$

---

\* For the purposes of the AP Exam, students will not be required to perform calculations using this equation; however, they must understand the underlying concepts and applications.

† For use with labs only (optional).

**1. YOUR NAME:**
(Print)     Last     First     M.I.

**SIGNATURE:**     DATE:   /   /

**HOME ADDRESS:**
(Print)     Number and Street

City     State     Zip Code

**PHONE NO. :**
(Print)

**IMPORTANT:** Please fill in these boxes exactly as shown on the back cover of your test book.

**5. YOUR NAME**

| First 4 letters of last name | | | | FIRST INIT | MID INIT |
|---|---|---|---|---|---|
| A | A | A | A | A | A |
| B | B | B | B | B | B |
| C | C | C | C | C | C |
| D | D | D | D | D | D |
| E | E | E | E | E | E |
| F | F | F | F | F | F |
| G | G | G | G | G | G |
| H | H | H | H | H | H |
| I | I | I | I | I | I |
| J | J | J | J | J | J |
| K | K | K | K | K | K |
| L | L | L | L | L | L |
| M | M | M | M | M | M |
| N | N | N | N | N | N |
| O | O | O | O | O | O |
| P | P | P | P | P | P |
| Q | Q | Q | Q | Q | Q |
| R | R | R | R | R | R |
| S | S | S | S | S | S |
| T | T | T | T | T | T |
| U | U | U | U | U | U |
| V | V | V | V | V | V |
| W | W | W | W | W | W |
| X | X | X | X | X | X |
| Y | Y | Y | Y | Y | Y |
| Z | Z | Z | Z | Z | Z |

**2. TEST FORM**

**3. TEST CODE**

**4. REGISTRATION NUMBER**

Test code letters: A B C D E F G (with digits 0–9)

Registration number columns: digits 0–9

**6. DATE OF BIRTH**

| Month | Day | Year |
|---|---|---|
| JAN | | |
| FEB | | |
| MAR | 0 0 | 0 0 |
| APR | 1 1 | 1 1 |
| MAY | 2 2 | 2 2 |
| JUN | 3 3 | 3 3 |
| JUL | 4 | 4 4 |
| AUG | 5 | 5 5 |
| SEP | 6 | 6 6 |
| OCT | 7 | 7 7 |
| NOV | 8 | 8 8 |
| DEC | 9 | 9 9 |

**7. SEX**
- MALE
- FEMALE

The Princeton Review
© The Princeton Review, Inc.
FORM NO. 00001-PR

## Section I

Start with number 1 for each new section.
If a section has fewer questions than answer spaces, leave the extra answer spaces blank.

1. A B C D
2. A B C D
3. A B C D
4. A B C D
5. A B C D
6. A B C D
7. A B C D
8. A B C D
9. A B C D
10. A B C D
11. A B C D
12. A B C D
13. A B C D
14. A B C D
15. A B C D
16. A B C D
17. A B C D
18. A B C D
19. A B C D
20. A B C D
21. A B C D
22. A B C D
23. A B C D
24. A B C D
25. A B C D
26. A B C D
27. A B C D
28. A B C D
29. A B C D
30. A B C D

31. A B C D
32. A B C D
33. A B C D
34. A B C D
35. A B C D
36. A B C D
37. A B C D
38. A B C D
39. A B C D
40. A B C D
41. A B C D
42. A B C D
43. A B C D
44. A B C D
45. A B C D
46. A B C D
47. A B C D
48. A B C D
49. A B C D
50. A B C D
51. A B C D
52. A B C D
53. A B C D
54. A B C D
55. A B C D
56. A B C D
57. A B C D
58. A B C D
59. A B C D
60. A B C D

61. A B C D
62. A B C D
63. A B C D
64. A B C D
65. A B C D
66. A B C D
67. A B C D
68. A B C D
69. A B C D
70. A B C D
71. A B C D
72. A B C D
73. A B C D
74. A B C D
75. A B C D
76. A B C D
77. A B C D
78. A B C D
79. A B C D
80. A B C D
81. A B C D
82. A B C D
83. A B C D
84. A B C D
85. A B C D
86. A B C D
87. A B C D
88. A B C D
89. A B C D
90. A B C D

91. A B C D
92. A B C D
93. A B C D
94. A B C D
95. A B C D
96. A B C D
97. A B C D
98. A B C D
99. A B C D
100. A B C D
101. A B C D
102. A B C D
103. A B C D
104. A B C D
105. A B C D
106. A B C D
107. A B C D
108. A B C D
109. A B C D
110. A B C D
111. A B C D
112. A B C D
113. A B C D
114. A B C D
115. A B C D
116. A B C D
117. A B C D
118. A B C D
119. A B C D
120. A B C D

**1. YOUR NAME:**

(Print)   Last   First   M.I.

SIGNATURE: _____   DATE: ___ / ___ / ___

**HOME ADDRESS:**
(Print)   Number and Street

City   State   Zip Code

**PHONE NO. :** _____
(Print)

IMPORTANT: Please fill in these boxes exactly as shown on the back cover of your test book.

**2. TEST FORM**

**6. DATE OF BIRTH**

| Month | Day | Year |
|---|---|---|
| JAN | | |
| FEB | | |
| MAR | 0 0 0 0 | |
| APR | 1 1 1 1 | |
| MAY | 2 2 2 2 | |
| JUN | 3 3 3 3 | |
| JUL | 4 4 4 | |
| AUG | 5 5 5 | |
| SEP | 6 6 6 | |
| OCT | 7 7 7 | |
| NOV | 8 8 8 | |
| DEC | 9 9 9 | |

**3. TEST CODE**     **4. REGISTRATION NUMBER**

**7. SEX**
- MALE
- FEMALE

**The Princeton Review**
© The Princeton Review, Inc.
FORM NO. 00001-PR

**5. YOUR NAME**

First 4 letters of last name | FIRST INIT | MID INIT

A B C D E F G H I J K L M N O P Q R S T U V W X Y Z

---

## Section 1

Start with number 1 for each new section.
If a section has fewer questions than answer spaces, leave the extra answer spaces blank.

1. A B C D
2. A B C D
3. A B C D
4. A B C D
5. A B C D
6. A B C D
7. A B C D
8. A B C D
9. A B C D
10. A B C D
11. A B C D
12. A B C D
13. A B C D
14. A B C D
15. A B C D
16. A B C D
17. A B C D
18. A B C D
19. A B C D
20. A B C D
21. A B C D
22. A B C D
23. A B C D
24. A B C D
25. A B C D
26. A B C D
27. A B C D
28. A B C D
29. A B C D
30. A B C D

31. A B C D
32. A B C D
33. A B C D
34. A B C D
35. A B C D
36. A B C D
37. A B C D
38. A B C D
39. A B C D
40. A B C D
41. A B C D
42. A B C D
43. A B C D
44. A B C D
45. A B C D
46. A B C D
47. A B C D
48. A B C D
49. A B C D
50. A B C D
51. A B C D
52. A B C D
53. A B C D
54. A B C D
55. A B C D
56. A B C D
57. A B C D
58. A B C D
59. A B C D
60. A B C D

61. A B C D
62. A B C D
63. A B C D
64. A B C D
65. A B C D
66. A B C D
67. A B C D
68. A B C D
69. A B C D
70. A B C D
71. A B C D
72. A B C D
73. A B C D
74. A B C D
75. A B C D
76. A B C D
77. A B C D
78. A B C D
79. A B C D
80. A B C D
81. A B C D
82. A B C D
83. A B C D
84. A B C D
85. A B C D
86. A B C D
87. A B C D
88. A B C D
89. A B C D
90. A B C D

91. A B C D
92. A B C D
93. A B C D
94. A B C D
95. A B C D
96. A B C D
97. A B C D
98. A B C D
99. A B C D
100. A B C D
101. A B C D
102. A B C D
103. A B C D
104. A B C D
105. A B C D
106. A B C D
107. A B C D
108. A B C D
109. A B C D
110. A B C D
111. A B C D
112. A B C D
113. A B C D
114. A B C D
115. A B C D
116. A B C D
117. A B C D
118. A B C D
119. A B C D
120. A B C D

# NOTES

# NOTES